THE SOLAREX GUIDE TO SOLAR ELECTRICITY

EDITED BY
ED ROBERTON

THE SOLAREX GUIDE TO SOLAR ELECTRICITY

**EDITED BY
ED ROBERTON**

**BY THE TECHNICAL
STAFF OF SOLAREX CORP.**

FIRST EDITION

FIRST PRINTING—APRIL 1979
SECOND PRINTING— OCTOBER 1979

Copyright © 1979 by Solarex Corporation

Printed in the United States of America

Reproduction or publication of the content in any manner, without express permission of the publisher, is prohibited. No liability is assumed with respect to the use of the information herein.

Preface

The idea of being able to get useful energy from the sun has been a dream of scientists for centuries. The sun is an inexhaustible source of energy, and sunlight is available all over the earth. What is needed to utilize the energy in sunlight is some efficient way to convert the radiant energy in sunlight into some form of energy that is easier to use, such as electrical energy. The silicon voltaic cell, or solar cell as it is commonly called, is just such a device.

The photovoltaic effect, where electricity is produced when certain materials are illuminated, is not new. It was first noted by E. Becqueral in 1839. Thus the photovoltaic cell is probably the first solid-state electronic device ever invented. It is certainly much older than the radio crystal detector which has often been called the first solid-state electronic device.

The utilization of the photovoltaic cell as an energy source has been very slow, primarily because of the abundance of hydrocarbon fuels such as coal, oil, and natural gas. The first practical use of the photovoltaic effect was in selenium cells that were, and still are, being used to measure light levels. A typical example is the light meter used in photography.

The first significant use of photovoltaic cells to produce electrical energy was in the space program. Satellites need a source of electrical energy that will last for a long time without any attention. All conventional batteries will run down after a period of use. The solar cell, however, will continue to deliver electric power as long as sunlight is available.

In recent years the development of solar cells has accelerated rapidly. This is due in part to the realization that our supply of hydrocarbon fuels, while large, is not inexhaustible. Another factor that has contributed to the development of solar power sources is the increasing concern about air and water pollution resulting from the burning of hydrocarbon fuels, and a possible hazard connected with the use of nuclear fuels.

Whereas the cost of conventional fuels has skyrocketed in recent years, the cost of solar cells has been declining steadily. The U.S. Department of Energy has a stated goal of reducing the cost of solar cells to $0.50 per peak watt by 1986. Even if this goal is not completely realized, the program will make the solar cell competitive with many other energy sources.

This book is intended for anyone interested in solar cells as a source of energy for any requirement, large or small. The first three chapters are intended as background material for readers who are not familiar with either solar energy or the photovoltaic effect. Chapters 4, 5, and 6 deal with practical solar cells, how to use them, and typical applications. A separate chapter describing a few solar cell projects is included for the experimenter who wishes to build either demonstration or practical energy sources. Chapter 8 covers commercially available accessories that make installation and use of a solar electric system easier.

The last chapter deals with the use of solar cells as light sensors rather than as power sources. A glossary of terms has been included for the benefit of those not familiar with the terminology that is springing up in connection with the application of solar electricity.

Technical Staff
Solarex Corporation

Contents

Introduction 9

1 A Little About Sunlight 11
Energy, Power & Power Density—How Much Energy From Sunlight?—The Units of Wavelenth.

2 The Photovoltaic Effect 19
Semiconductors—The Photovoltaic Cell.

3 The Silicon Solar Cell 25
Current Versus Voltage Curves—Maximum Power—Output Power Versus Sunlight—The Solar Cell as an Energy Source—Spectral Response—Energy Conversion Efficiency.

4 Practical Solar Electric Generators 35
Construction & Efficiency—Temperature Influence—Solar Cell Specifications—Series & Parallel Connections—Solar Panels—High Density Solar Panels—Concentrator Cells—Solar Microgenerators.

5 How to Use Solar Cells 59
Available Sunlight—Load Requirements—Intermittent Loads—System Calculations—Batteries—Types of Batteries—Battery Specifications—The Complete System.

6 Applications of Solar Cells 75
Radio & Television—Fiber Optics—Agricultural Applications—The Construction Industry—Remote Areas—Solar Electric Cathodic Protection—Applications of Concentrator Cells—Military Applications—Microgenerator Applications—Solar Powered Flashlight—Solar Powered Attic Fan—Solar Power for Recreation Future Applications of Solar Electricity—Two-Way Communications.

7 Solar Electric Projects... 109
Demonstration Projects—Components—Direct Solar Electric Power—Battery Charging—Increasing Voltage From Solar Cells.

8 Accessories... 125
Charge Controller—Measurement Equipment—Trackers.

9 Nonpower Photovoltaic Cell Applications........................... 131
Light Meter—Light Operated Relays.

Glossary of Solar Electric Terms 137

Index.. 141

Introduction

When solar cells were invented in the early 1950's, their use for terrestrial applications was considered. At that time the cost of electricity and energy producing means was still declining, and the newly-invented solar electricity was not even in the ballpark. It might have been forgotten if the space program had not come about.

What was expensive for earth use—when oil, gas, batteries, etc. were abundant—was inexpensive for space use. The U.S. space program then brought solar cells out of the laboratory into limited production, and they became the power source for the satellites. Because of the required high reliability and the premium on size and weight, space solar cells became rather expensive and efficiency became the key word.

Solar cell efficiency improved slowly for 10 years until an abrupt improvement occurred in 1972 and a high efficiency solar cell was developed.

Although solar cells proved to be an extremely reliable and simple source of electricity for space use, due to their cost, only a small number of space reject cells were used for terrestrial applications as late as 1972. The primary reason was that solar cell technology was not applied to production of inexpensive solar cells for terrestrial use. In 1972 it was considered that the cost of solar electricity was over $100/watt compared to about $1/watt or less for conventional electricity.

Dr. Joseph Lindmayer, known for his work in solid state physics and who invented the high efficiency solar cell, and Dr. Peter Varadi, who had broad experience in materials technology and business administration, recognized that solar electricity was feasible and would be one of the most important energy sources of the future. At that time no serious effort was made by any company to develop and mass produce solar cells and generators for terrestrial use only.

With this in mind, Solarex Corporation was formed in early 1973 with the purpose of devoting its entire effort to solar energy and to advancing the state of art by progressive research and development and manufacturing.

Solarex Corporation is a leader in the development of terrestrial solar cells to convert light to electricity. It has total manufacturing capability to produce efficient solar cells and to integrate them into solar panels. Solarex builds standard solar panels (Unipanels®) and assembles these building blocks into larger solar-electric generators.

Solarex is equipped to develop and custom-design any type of solar-electric cell, panel or system. Solarex products will always incorporate the most advanced concepts in solar-electric conversion. In 1973 Solarex developed its terrestrial solar cell which, even by its appearance, can be easily distinguished from other solar cells.

Solarex is located in the suburbs of Washington, D.C. (20 miles from the White House) in a modern building, selected and designed to provide a research and development and production facility as well as a capability for undisturbed environmental testing of solar conversion equipment. Engineering data obtained from this test facility is being used in the design of solar-electric systems.

Chapter 1
A Little About Sunlight

Sunlight is one of those things that most of us take for granted. We know in a vague sort of way that we can get energy from the sun, but that's about the extent of our knowledge. The fact is that all energy on earth originally came from the sun. All of our hydrocarbon fuels such as coal, oil, and natural gas, were originally produced by the action of sunlight on vegetation.

In order to understand how we can use solar cells to get electrical energy from sunlight, we should know a little bit about the nature of light and the units that are used to measure it. Light is a form of electromagnetic energy—just like a radio or a television signal. The only difference is that the wavelength of light is very much shorter than that of radio waves. The color of light is determined by its wavelength. The longest light waves that we can detect visually are red in color and the shortest are violet.

ENERGY, POWER & POWER DENSITY

How do we measure the energy that reaches the earth from the sun? What units do we use to express it? From elementary physics we learn that energy is the capacity to do work. It is measured in such units as the gram-calorie, the

Table 1-1. Conversion Factors For Energy

MULTIPLY → BY TO GET ↓	WATT-HOURS	KILOWATT-HOURS	JOULES (WATT-SECONDS)	KILOGRAM-CALORIES
WATT-HOURS	1	10^3	2.778×10^{-4}	1.163
KILOWATT-HOURS	10^{-3}	1	2.778×10^{-7}	1.163×10^{-3}
JOULES (WATT-SECONDS)	3.6×10^3	3.6×10^6	1	4.187×10^3
KILOGRAM-CALORIES	8.599×10^{-1}	8.599×10^2	2.388×10^{-5}	1

joule, or the kilowatt-hour. The ordinary household electricity meter measures electrical energy in kilowatt-hours.

An inconceivable amount of energy is radiated by the sun. Only a very small fraction of this energy—about one-half of one billionth—ever reaches the earth. This small fraction, however, is still a tremendous amount of energy. It has been estimated that every year about 745 quadrillion kilowatt-hours of energy reach the earth from the sun. It is understandable why scientists have sought for years for ways to harness this energy.

Although energy is the more fundamental concept, in practical applications we are usually more interested in *power*, which is the rate at which energy is generated or used. Power is measured in watts or kilowatts.

The amount of power that we can get from any device that harnesses sunlight depends on the amount of sunlight that it intercepts. A device that has a larger area will intercept more power from sunlight than a smaller device. For this reason we often speak of sunlight in terms of *power density*. Power density is simply the amount of power in a given area. We usually express power density in terms of milliwatts per square centimeter (mW/cm^2) or kilowatts per square meter (kW/m^2).

Historically, most of the measurements of the amount of sunlight reaching the various parts of the earth have been made by people interested in forecasting weather. Not surprisingly, they have their own unit of measurement for the purpose. It is the *langley*. One langley is equal to 11.62

Table 1-2. Conversion Factors For Power

MULTIPLY → BY TO GET ↓	MILLIWATTS	WATTS	KILOWATTS
MILLIWATTS	1	10^3	10^6
WATTS	10^{-3}	1	10^3
KILOWATTS	10^{-6}	10^{-3}	1

watt-hours per square meter. In Chapter 5, we will show how to use a more convenient unit, the number of *peak sun hours*.

In the literature about solar energy you will find these and other units used to measure energy, power, and power density. Tables 1-1, 1-2, and 1-3 show how to convert from one such unit to another. Using a small pocket calculator the conversion is very easy.

HOW MUCH ENERGY FROM SUNLIGHT?

The power density of sunlight reaching the outside of the earth's atmosphere is about 136 mW/cm^2. About one-third of

Table 1-3. Conversion Factors For Power Density

MULTIPLY → BY TO GET ↓	mW/cm^2	W/m^2	kW/m^2
mW/cm^2	1	10^{-1}	10^2
W/m^2	10	1	10^3
kW/m^2	10^{-2}	10^{-3}	1

this energy is scattered while passing through the earth's atmosphere. Thus on a clear day at noontime the power density of sunlight is about 100 mW/cm². This amounts to about 1 kW/m² or in more familiar units about 1 kilowatt per square yard.

Of course the actual power density at the surface of the earth depends on many factors. The figures quoted above are for a clear day at noontime. Early in the morning and late in the afternoon, when the sun is lower in the sky, the amount of energy received is lower, as it is on cloudy or rainy days.

The actual amount of power that we can recover from sunlight depends not only on the amount of sunlight that reaches a given area in a year, but also on the efficiency of our energy conversion device. These factors will be treated in detail in Chapter 5. We can, however, get some idea of the amount of energy that we are talking about from the fact the solar energy that is intercepted by a small tennis court would supply the energy needs of an average household. The solar energy that reaches less than one percent of the Sahara desert is more than all of the electrical energy used by all of the nations of the world.

THE UNITS OF WAVELENGTH

We mentioned earlier that light is actually a form of electromagnetic energy. The wavelength of visible light is very short as compared to that of a radio wave and therefore the usual units of measurement of wavelength, such as the

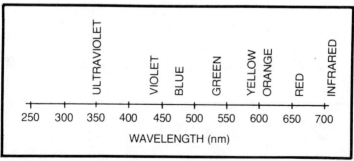

Fig. 1-1. Color of light as a function of wavelength. Units are in nanometers.

Table 1-4. Conversion Factors for Wavelengths

MULTIPLY BY / TO GET	MICRONS	ANGSTROMS	METERS	NANOMETERS
MICRONS	1	10^{-4}	10^6	10^{-3}
ANGSTROMS	10^4	1	10^{10}	10
METERS	10^{-6}	10^{-10}	1	10^{-9}
NANOMETERS	10^3	10^{-1}	10^9	1

meter, are inconveniently large. A suitable unit for the measurement of the wavelength of light is the *nanometer*. A nanometer is equal to a billionth of a meter, or in mathematical language, 10^{-9} meter. In some of the older literature on light the nanometer is called a millimicron. A micron is a millionth of a meter. Another name for the micron is micrometer.

Figure 1-1 shows the relationship between the wavelength and its color. Table 1-4 can be used to convert between nanometers and other units that are sometimes used to express the wavelength of light. Figure 1-2 shows the spectral distribution, that is, the amount of each color in ordinary daylight.

For many years, the nature of light was a subject of hot debate by physicists. At the time of Sir Isaac Newton, about

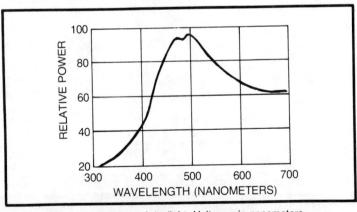

Fig. 1-2. Spectral distribution of daylight. Units are in nanometers.

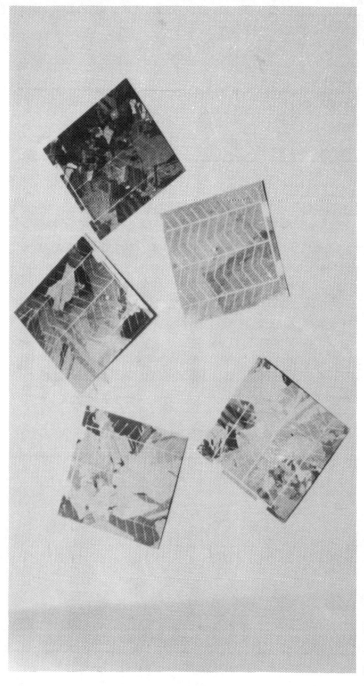

Fig. 1-3. The single-crystal semiconductor-grade silicon from which most solar cells are made is extremely expensive. Cells fabricated from relatively impure semicrystalline silicon, such as shown here, offer great potential for cutting solar-cell prices.

the year 1666, light was considered to consist of tiny particles or corpuscles. Later, in 1865, Maxwell showed that light was wavelike in nature. Still later, in 1905, Albert Einstein showed that light was corpuscular in nature. Actually, there is evidence to support both theories. To the layman, this situation is apt to be confusing, but the easy approach is to use whichever point of view works best in a particular situation.

In most applications the wave theory of light will work perfectly well. We do not need to get involved in particle physics. The exception is when we try to understand what goes on inside a solar cell. Here, we can best describe what happens by considering light to consist of small bundles of energy called *photons*. The photon is the smallest particle of light that can exist physically.

Chapter 2
The Photovoltaic Effect

The silicon solar cell is actually a *photovoltaic* cell. The word photovoltaic means the direct conversion of light into electricity. The photovoltaic cell is a semiconductor device. In order to understand how it works we will first review the basic principles of semiconductors.

SEMICONDUCTORS

A semiconductor material, such as silicon or germanium, is a material that is classified somewhere between being a good conductor and being a good insulator. Silicon is one of the most commonly used semiconductors. There is not likely to be a shortage of it because about twenty-five percent of the earth's crust is made up of silicon. Silicon usually occurs as a compound such as the silicon dioxide that makes up ordinary sand. It can be reduced to a pure metallic form by a reduction process.

After silicon has been refined and purified it can be formed into a crystalline structure. Each atom in the outer ring of the silicon atom has a definite place in the crystal structure. Thus there are not any free electrons in a pure silicon crystal. As a result it is a rather poor conductor of electricity because a lot of force is required to draw the electrons out of their places in the crystal lattice.

To make silicon that is useful in electronic devices such as diodes, transistors, or solar cells, the material is changed by adding minute amounts of other elements. This process is called *doping* of the silicon.

When a very small amount of phosphorus is added to silicon while the crystal is being formed, there will be some free electrons in the material which do not have definite places in the crystal structure. These free electrons can move about easily and are called negative, or N-type charge carriers. Silicon that is doped with phosphorus is called N-type silicon.

If, instead of adding phosphorus, we add boron to the silicon while the crystal is being formed, we have a completely different situation. There are no free electrons. In fact there are places in the crystal where there would be electrons in pure silicon. These voids in the crystal are called *holes*. These holes can move easily from one place in the crystal structure to another. What really happens is that one of the electrons in the crystal structure will fill one of the holes, leaving another hole in the place that it came from. Inasmuch as the holes actually move by electrons filling them, each hole will have a positive charge that is equal and opposite to the negative charge of an electron. Silicon doped with boron is called P-type, or positive type silicon.

Figure 2-1 shows the extra holes in P-type material and the extra electrons in N-type material. As we stated earlier, the holes have a positive charge and the electrons have a negative charge. Note, however, that this does not mean that the material itself has an electric charge. Although the electrons and holes have electric charges, there are an equal number of positive and negative charges in the nucleus of the atoms of the material that are not free to move around in the material. These charges cancel out the charges of the holes and electrons so that the material itself is electrically neutral. Figure 2-2 shows how P-type and N-type silicon can be combined to form a semiconductor junction.

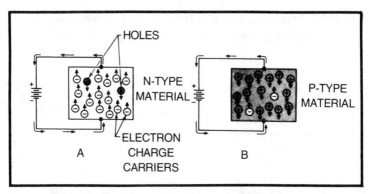

Fig. 2-1. Current flow in semiconductor materials. (A) N-type material. Majority of current flow is electron movement. (B) P-type material. Majority of current flow is hole movement.

To understand how light interacts with a semiconductor material, we will make use of both the corpuscular and wave theories of light. First let's think of light as consisting of particles called photons. When a photon penetrates a semiconductor material such as silicon, it will force an electron out of its place in the crystal structure. This forms what we call a *hole-electron* pair. The electron has a negative charge, and the hole that it leaves will have a positive charge.

If we don't make any other provisions, this hole-electron pair will have a very short life indeed. In a period of about a millionth of a second, the electron will move right back into the hole in the crystal that it came from. This process is called *recombination* of the hole and the electron.

The depth to which a photon will penetrate the silicon depends on its energy. This is most easily explained by going back to the wave theory of light. Light of a longer wavelength will produce photons having more energy. In practical terms this means that the degree to which light will penetrate into the silicon depends on its wavelength. In general, light having a longer wavelength will penetrate more deeply into the silicon.

THE PHOTOVOLTAIC CELL

Figure 2-3 shows a sketch of a photovoltaic cell. The base material at the right of the figure is made of P-type silicon.

After the P-type material is formed by adding boron doping, a thin layer of N-type material is formed by changing the doping to phosphorus.

The N-type layer is very thin, and although it appears opaque to the eye, sunlight will penetrate rather deeply into it. In fact, the light actually penetrates the junction between the

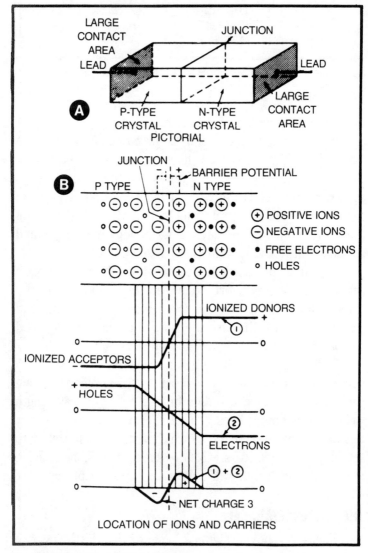

Fig. 2-2. Formation of a PN junction.

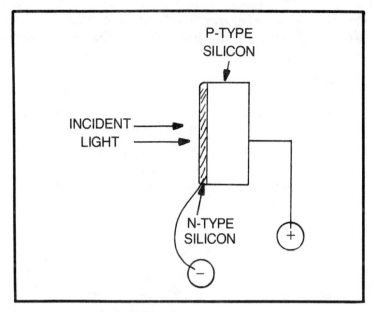

Fig. 2-3. Elementary silicon photovoltaic cell.

N-type and P-type material. This is where the hole-electron pairs that we talked about earlier are formed. The electric field that exists at the junction will prevent the holes and electrons from recombining, with the result that we can now use the

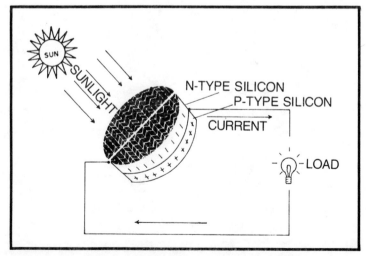

Fig. 2-4. Solar cell connected to a load.

device as a source of energy. The N-type material will be the negative pole and the P-type material will be the positive pole of what amounts to a little electric generator that gets its energy from sunlight.

Figure 2-4 shows how a solar cell is connected to a load. The photovoltaic action of the sunlight on the solar cell will cause current to flow through the load. In the following chapter we will see how the amount of electricity produced by the cell is related to the amount of light, and how the load can be adjusted to get the maximum amount of energy from the cell.

Chapter 3
The Silicon Solar Cell

In the preceding chapter we saw how the photovoltaic effect in silicon can be used to generate electricity directly from sunlight. In this chapter we will take a detailed look at the electrical characteristics of a solar cell utilizing this effect. In particular we will see how the voltage and current produced by a solar cell vary with different load conditions and how the output varies with the amount of available light.

CURRENT VERSUS VOLTAGE CURVES

Whenever we use any source of electric energy, it is helpful to know just how the terminal voltage and current will change as the load is varied. The solar cell is a nonlinear device, so it isn't convenient to try to use a mathematical expression for this characteristic. Instead, we will use a family of easily understood curves on a graph.

One way that we can plot such a set of curves is to connect the solar cell in the circuit shown in Fig. 3-1 and vary the load resistance over a wide range while noting the values of voltage and current indicated on the meters in the circuit.

Suppose that we have a silicon solar cell connected in the circuit of Fig. 3-1 and it is subjected to an amount of light

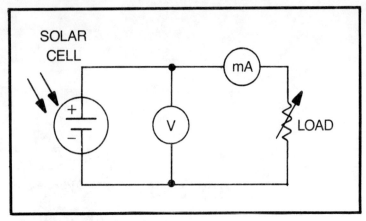

Fig. 3-1. Circuit for plotting voltage-current curves.

approximately equal to the maximum value of sunlight at the surface of the earth—about 100 mW/cm^2. To start, suppose that we make the load resistance so high that for all practical purposes we can consider it to be an open circuit. Naturally, the current indicated on the milliammeter will be zero because no current can flow into an open circuit. With the silicon cell, the terminal voltage under this condition will be about 0.570 volts. This point is labeled A in Fig. 3-2. It is also labeled as V_{OC}—the open-circuit voltage.

Now let us assume that we slowly decrease the load resistance. We will find that the terminal voltage drops very slowly, but that the current increases rather rapidly. This continues until we reach point B on the curve. At this point the voltage has dropped to about 0.45 volts and the current has increased to about 780 milliamperes. The shape of the curve changes rather drastically at this point, which is often referred to as the knee of the curve.

If we continue to decrease the load resistance, the voltage will continue to drop, but the current will remain nearly constant. When the load resistance reaches zero, the voltage will, of course, be zero because we can't have a voltage across a short circuit. This is labeled as point C on the curve. It is also labeled as I_{SC} because it shows the short-circuit current of the cell.

MAXIMUM POWER

Now that we have a curve that shows the voltage-current characteristic of the solar cell, the question naturally arises as to just where on the curve we should try to operate the cell. We can, of course, establish the operating point by properly adjusting the load resistance. In fact, the load resistance that will cause the cell to operate at any point on the curve can be found by dividing the voltage by the current at that particular point on the curve.

In most practical applications we will want to operate the cell in such a way that it will deliver the maximum amount of power to the load. Figure 3-3 shows the same graph that we showed in Fig. 3-2, but now we have added another curve.

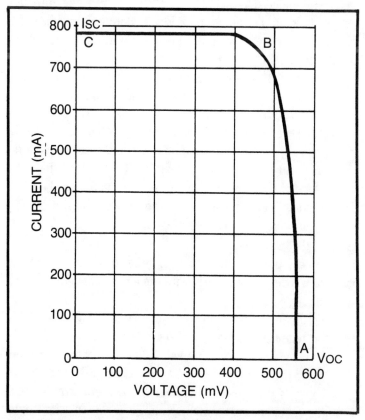

Fig. 3-2. Voltage-current curve for a solar cell.

27

The dashed line and the scale at the right of the graph show the amount of power that the cell will deliver to the load when it is operated at any point on the curve. Intuitively, we know that can't get much power into either an open or a short circuit. We expect that amount of power that we can get to be maximum when we operate the cell somewhere between these two extremes.

Interestingly, the power that is delivered to the load will be maximum when we operate the cell at about the knee in the characteristic curve.

OUTPUT POWER VERSUS SUNLIGHT

The curves of Fig. 3-2 and 3-3 were obtained when the cell was subjected to the maximum value of sunlight. Although this is an ideal operating condition, maximum sunlight is not always available, so we must know how the cell will behave when the level of light is reduced.

Figure 3-4 shows a family of curves similar to those of Figs. 3-2 and 3-3, but for different values of light reaching the solar cell. The top curve of Fig. 3-4 is almost exactly the same curve that we studied in Figs. 3-2 and 3-3. The next curve on the graph represents what happens when the power from the sunlight is cut in half. Note that to the left of the knee the curve has the same shape as the curve above it and the one below it. In this portion of the operating range the terminal voltage is not significantly affected by the amount of light. It is the current that changes. Inasmuch as the voltage doesn't change very much, power will be proportional to current, and the current will decrease linearly with the amount of light reaching the cell.

To the left of the knees of the curves in Fig. 3-4 we see that the terminal voltage does to some extent depend on the amount of light. At very low levels, about 8% of the maximum sun level, the voltage decreases noticeably. Nevertheless, over most of the operating range, the current from the cell changes much more than the voltage when the light level changes.

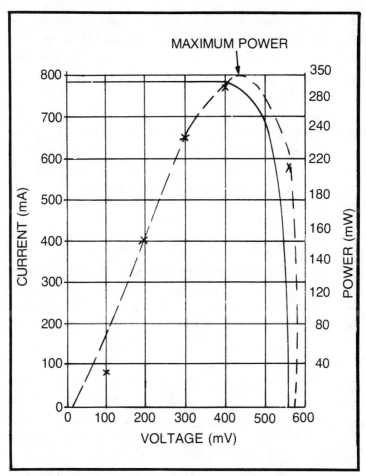

Fig. 3-3. Voltage-current curve of solar cell with power plotted also.

This has an important implication in many practical applications of solar cells. The fact that the curves have the same shape except at very low light levels means that we will get nearly the maximum amount of power at all except the lowest light levels if we operate the cell at the knee of the curve. Thus, in many applications, there is no need to change the load as the amount of light varies.

We can also draw a few more conclusions from the family of curves shown in Fig. 3-4. In the first place, we note that except at low light levels the voltage from the cell doesn't change very much with the amount of light. From this we might guess that the voltage from a cell wouldn't change very much as the cell was made larger or smaller. This is indeed true. The voltage produced by a solar cell is roughly 0.45 volts regardless of the area of the cell.

We also note that the current produced by the cell varies greatly with the amount of light reaching the cell. This would lead us to suspect that a cell with a larger area would produce more current. This suspicion is correct. The amount of current from a solar cell increases linearly with both the area and the amount of light.

THE SOLAR CELL AS AN ENERGY SOURCE

The voltage-current curves of a solar cell, such as those shown in Fig. 3-4, completely describe its electrical behavior. From these curves we see that the solar cell is not much like other voltage sources with which we may be familiar. A battery, for example, behaves very much like a constant voltage source except at very high load currents. The current from a battery will increase as the load resistance decreases, but the terminal voltage will not change very much until the load current becomes very high. We can't operate a battery into a short circuit. The current would become excessive and the battery would be destroyed.

By contrast, the solar cell acts very much like a constant current source over most of its operating range. Even if the load resistance were to become very low, the current would not become excessive. This feature makes the solar cell ideal for unattended operation. We don't have to keep an eye on the load current because the cell will not destroy itself.

SPECTRAL RESPONSE

Another important characteristic of any photoelectric device is what is called its spectral response. This is its relative response to light of different wavelengths. Figure 3-5 shows the spectral response of a silicon solar cell. The horizontal axis

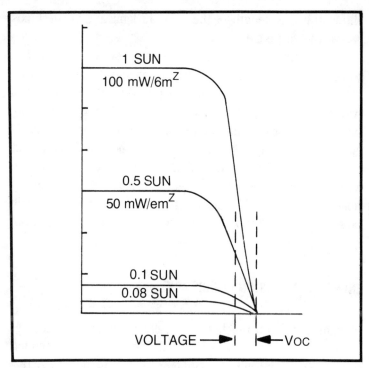

Fig. 3-4. Solar cell output of different amounts of sunlight.

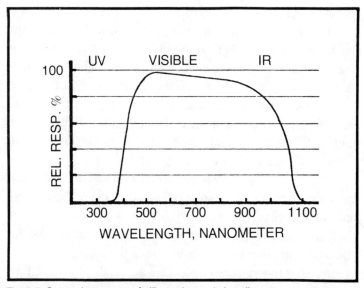

Fig. 3-5. Spectral response of silicon photovoltaic cell.

shows the wavelength of the incident light and vertical axis shows the response.

Of course, we don't have any control over the wavelength of sunlight, but the curves show that the silicon solar cell has a relatively high response to a broad range of wavelengths ranging from the ultraviolet, through the visible portion of the spectrum, into the infrared. This means that the solar cell will produce electrical energy under a wide variety of conditions. It will respond well on cloudy days and will produce a useable output when illuminated by ordinary incandescent or flourescent lamps. This makes the silicon solar cell useful in photoelectric applications other than generating power from sunlight.

ENERGY CONVERSION EFFICIENCY

So far, we have talked about the amount of power in the sunlight that might reach a solar cell and about the electrical power that we can get out of the cell. We haven't said anything about the energy conversion efficiency. This is the ratio of the power that we can get out of a cell to the power in the sunlight that it receives.

It turns out that there is a theoretical limit to the maximum efficiency that can be obtained with a silicon solar cell of the type that we have been discussing. This theoretical limit of energy conversion efficiency turns out to be about twenty-five percent. This sounds like a very low figure until we realize that energy conversion efficiencies in this order are common. The ordinary automobile engine has an energy converison efficiency of only about twenty percent.

There are many practical considerations that determine just how efficient a given solar cell will be. One of the connections to the cell is made to the thin layer of N-type material on the face of the cell. If the metallic conductors used to make this connection are too large, they will block the sunlight from part of the cell. If the metallic conductors are made too small, they will have a high electrical resistance.

Fig. 3-6. A 2000-watt array of solar panels at Washington, D.C., for a national Sun Day celebration powers a water fountain and the world's first photovoltaic band.

When individual cells are combined into an array, their geometrical configuration will determine whether or not they make maximum use of the available area. The efficiencies that can be expected from practical solar cells will be discussed in the following chapter.

Chapter 4
Practical Solar Electric Generators

The basic element of a solar electric generator is the silicon solar cell. These cells are available in a wide variety of sizes and shapes, each of which is suitable for many applications. Figure 4-1 shows a few of the solar cell configurations that are now available. The most popular shapes are rectangles, circles, and segments of circles.

When the application calls for a considerable amount of power, more than a single cell must be used in the generator. For such applications many cells are combined into assemblies called solar panels. When still more power is required, several panels can be combined into large arrays.

At the other extreme, when only a very small amount of power is required, several very small cells are combined into assemblies called Microgenerators® which can be used to supply power to small electronic devices such as watches, calculators, etc.

Figure 4-2 shows a sketch of a solar cell. The manufacturing process starts out with nearly one hundred percent pure silicon material. To form the bottom layer of the cell, boron is added in minute amounts while the silicon crystal is being formed. This makes the bottom layer of the cell P-type ma-

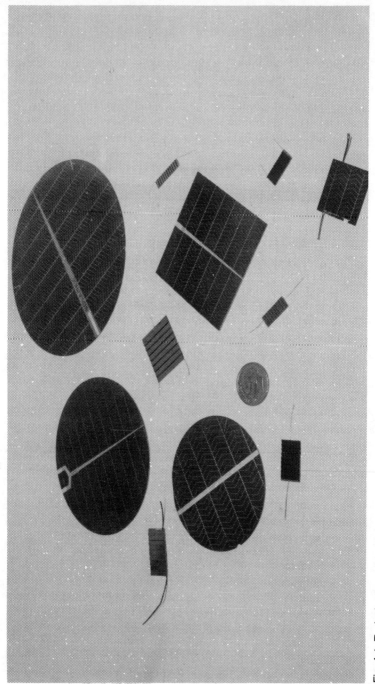

Fig. 4-1. Typical solar cell configuration now available.

terial. After the P-type layer has been formed, phosphorus is added to form the thin layer of N-type material, thus forming the PN junction which is essential for the operation of the cell. Metallic connections are provided to both the P-type and N-type layers.

CONSTRUCTION & EFFICIENCY

We noted earlier that the theoretical limit of efficiency of a photovoltaic cell is in the order of twenty-five percent. The actual efficiency that can be obtained in practice is somewhat lower than the theoretical limit and depends to a great extent on the manufacturing process and materials used to make the cell.

The efficiency of the cell is quite important because the higher the efficiency, the fewer cells will be required in a given application. For example, a three-inch circular cell with an efficiency of twelve percent will provide 0.547 watts at 1.2 amperes. A similar three-inch cell with an efficiency of ten percent would only provide a power of 0.456 watts at 1.0 amperes.

The efficiency of a solar cell is the ratio of the power output to the power input. It can be expressed by the equation:

$$\text{Efficiency Percentage} = \frac{\text{Power Output}}{\text{Power Input}}$$

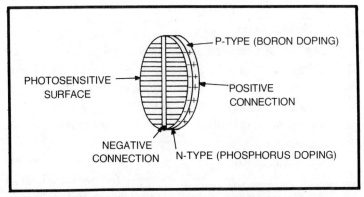

Fig. 4-2. Construction of solar cell.

Many factors influence the efficiency of a practical solar cell. In order for the cell to be able to supply power to an external circuit we must have metallic connections to both the P-type and the N-type layers of the cell. The connection to the P-type bottom layer isn't too difficult. A metallic plate at the bottom of the cell serves this purpose. The connection to the top N-type layer presents more of a problem.

The N-type layer is intentionally very thin so that light can penetrate through to the PN junction where the photovoltaic action takes place. Making a good connection to such a thin layer requires a precision process. The situation is further complicated by the fact that the metallic material used to make the connection will block some of the light that would otherwise penetrate to the junction. This blocking will reduce the efficiency of the cell.

The metallization used to make the connection must resist corrosion if the cell is to have a long life under outdoor conditions.

Another factor that can reduce the efficiency of a solar cell is reflection from its surface. Sunlight that is reflected from the surface will not penetrate to the junction, and will not be converted into electrical energy. Practical cells have a very thin antireflection coating similar to that used on photographic lenses. This layer reduces reflection without significantly blocking any of the incident light from the cell.

TEMPERATURE INFLUENCE

The silicon solar cell is normally rated for use between the temperatures of −65 degrees to +125 degrees Celsius (−85 degrees to +257 degrees Fahrenheit). The cell will withstand a temperature of up to +250 degrees Celsuis for a period of not more than 30 minutes, and even a temperature of 300 degrees Celsius for shorter periods of time.

Solar cells perform very well at extremely low temperatures. It is not uncommon to find a lower temperature limit as low as −100 degrees Celsius (−148 degrees Fahrenheit).

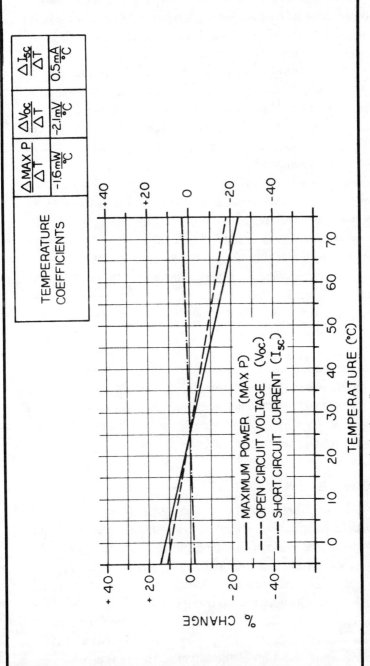

Fig. 4-3. Temperature characteristics of a typical solar cell.

The temperature of a solar cell influences the amount of power that it will deliver for a given amount of illumination. As the temperature of the cell increases, the terminal voltage will decrease at a rate of about 2 millivots per degree Celsius. This decrease in voltage is partially compensated by the fact that the current will increase at a rate of about 0.5 milliampere per degree Celsius. The compensation is not complete, however, and the net effect is that as the temperature of the cell increases the available power will decrease at a rate of about 0.3 percent per degree Celsius.

Figure 4-3 shows a plot of open-circuit voltage, short-circuit current, and maximum power output as functions of temperature. Inasmuch as the cell is rated for operation at a temperature of 25 degrees Celsius, the changes shown in the graph are zero at that temperature.

The curve of most interest is usually the maximum power output shown by the solid line. This curve shows that at temperatures below 25 degrees Celsius the power output actually increases. At higher temperatures, the power output falls off.

Instead of using the curves of Fig. 4-3 the effect of temperature can be found from the simple formulas in Fig. 4-4. Here a separate formula is used for temperatures above and below 25 degrees Celsius because the specification of the cell applies at this temperature.

SOLAR CELL SPECIFICATIONS

In selecting a solar cell or combination of cells for any particular application, we need to compare the specifications of the cell we select with the requirements of the application. A typical list of solar cell specifications is given in Fig. 4-5.

The first consideration in fitting a cell to a particular application is its geometrical configuration. The cells specified in Fig. 4-5 are separated into circular, hexagonal, and rectangular shapes. Of course, the physical size is important and the specification sheet gives the physical size of the cell.

$$E_{OUT} = E_{REF} [1 - 0.002 (T - 25)]$$

E_{OUT} = OUTPUT OF CELL IN VOLTS
E_{REF} = OUTPUT OF CELL IN VOLTS AT 25°C
T = TEMPERATURE IN DEGREES CELSIUS

$$I_{OUT} = I_{REF} [1 + 0.025A (T - 25)]$$

I_{OUT} = OUTPUT OF CELL IN MILLIAMPERES
I_{REF} = OUTPUT OF CELL IN MILLIAMPERES AT 25°C
A = AREA OF CELL IN SQUARE CENTIMETERS
T = TEMPERATURE IN DEGREES CELSIUS

Fig. 4-4. Equations for calculating solar cell output voltage and current at various temperatures.

The electrical specifications of importance are the output current and the efficiency. It is usually not necessary to state the voltage output of the cell, because the current output is specified at a particular voltage. For example, in Fig. 4-5, the output currents of the various cells are given at a voltage of 0.45 volts.

A detailed analysis of the specifications given in Fig. 4-5 will reveal that cells having approximately the same area and efficiency may have slightly different output currents. This is due to the fact that the dome on the surface area of the cell is shaded by the bus bar that is used to make the electrical connection. Different bus bar configurations result in different amounts of shading.

Other specifications that are of interest are the environmental specifications. These include the temperature range over which the cell will operate and its resistance to various environmental influences such as humidity. For example, the specification might contain the statement, "Humidity tested

Typical Solar Cell Specifications

	TYPE NO.	SIZE		CURRENT (min.)	EFFICIENCY
		Inches	mm	(mA at 0.45 V)	(min.)[3]
Circular Cells	20T202	2" dia.	50	500	11
	25T203	2¼" dia.	56	550	10
	25T227	2¼" dia.	56	670	12
	25T234	2¼" dia.	56	775	14
	12T236	2¼" (1)	56	270	10
	12T237	2¼" (1)	56	330	12
	6T204	2¼" (2)	56	130	10
	6T228	2¼" (2)	56	160	12
	44T210	3" dia.	75	1000	10
	44T229	3" dia.	75	1200	12
	22T238	3" (1)	75	500	10
	22T239	3" (1)	75	600	12
	11T240	3" (2)	75	280	12
	81T233	4" dia.	100	2150	12
Hexagonal Cell	38T230	(4)	(4)	1000	12
Rectangular Cells	012T225	0.2x0.1	5x2.5	1.5	10
	012T226	0.1x0.2	2.5x5	2	10
	025T220	0.2x0.2	5x5	4	10
	05T217	0.2x0.4	5x10	11	10
	05T219	0.4x0.2	10x5	8	10
	1T216	0.4x0.4	10x10	22	10
	1T218	0.8x0.2	20x5	19	10
	2T207	0.4x0.8	10x20	36	8
	2T208	0.8x0.4	20x10	32	8
	2T205	0.4x0.8	10x20	45	10
	2T209	0.8x0.4	20x10	42	10
	4T221	0.8x0.8	20x20	70	8
	4T206	0.8x0.8	20x20	90	10
	4T222	0.8x0.8	20x20	105	12
	4T223	0.8x0.8	20x20	125	14
	4T235	0.8x0.8	20x20	140	16
	8T211	1.6x0.8	40x20	180	10
	8T231	1.6x0.8	40x20	210	12
	12T212	2.4x0.8	60x20	265	10
	12T232	2.4x0.8	60x20	320	12

Notes to table: Cells with the same nominal area and efficiency may have slightly different current output due to the surface area shaded by various bus bar configurations. Bus bar location on rectangular cells is on the first dimension indicated, i.e., the bus bar on a 100mm × 20mm cell is located on the 10mm side.

Open circuit voltage (VOC) of all cells is 0.55 volts minimum.

(1) These cells are semicircular segments of cells of the indicated diameter.
(2) These cells are quadrants of circular cells of the indicated diameter.
(3) These figures show the guaranteed minimum efficiency at 0.45 V. The average efficiency of all cell types at actual maximum power point exceeds the number shown by at least one percentage point (e.g., a cell with 10% indicated minimum efficiency will be at least 11% efficient).
(4) This hexagonal cell measures 2½" between parallel sides.

Fig. 4-5. Solarex cells are supplied with tabs attached to the positive and negative terminals and coated with transparent protective coatings. Cells may be purchased with color coded wires, also with transparent protective coating (denoted by suffix/W). Solar cells are also available uncoated for assembly without tabs or wires (denoted by suffix/Z). Solar cell efficiency is measured with transparent protective coating on the cell's active surface.

for longer than seven days at 90 percent relative humidity and 70 degrees Celsius."

SERIES & PARALLEL CONNECTIONS

The power output of a typical solar cell is about the same as that from an ordinary flashlight battery. The voltage output of a single cell is about 0.45 volts under full sun conditions. The amount of current produced by a single cell depends on its area. We can get an idea of the current output from the fact that the full sun output current of a 4-inch cell is about 2 amperes. Thus in many applications it is necessary to connect several cells together to form a generator that will supply the necessary voltage and current.

Solar cells can be connected in series and in parallel just like other cells such as flashlight cells. As shown in Fig. 4-6A, when several cells are connected in series, the total voltage from the assembly will be the sum of the voltages from the individual cells. The output current will be the same as that from a single cell.

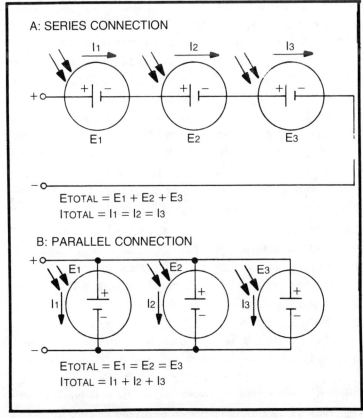

Fig. 4-6. Solar cells connected in (A) series and (B) parallel.

When cells are connected in parallel, as shown in Fig. 4-6B, the total current will be the sum of the currents from the individual cells, but the voltage will be the same as that from a single cell.

In most applications cells are connected in both series and parallel as shown in Fig. 4-7. Here the number of series groups of cells will determine the voltage and the number of parallel cells in each group will determine the current that can be delivered to a load.

SOLAR PANELS

When several solar cells are connected together in series parallel to obtain the desired voltage and current ratings the assembly is called a solar panel, Unipanel®, or a Solar Energizer®. Figure 4-8 shows a typical solar panel. This particular panel will deliver up to 10 watts of electrical power at 7 volts.

The actual construction of a solar panel depends to a large extent on the conditions to which it will be subjected in any particular application. In the Unipanel type of construction the cells are mounted on a rigid random-fiber polyester board. The cells themselves are usually encapsulated in a highly transparent silicone rubber. Although this encapsulation provides a great deal of protection to the surface of the cells, some severe environmental conditions require additional protective coverings in addition to the silicone rubber. Coatings of glass, Plexiglas, or Lexan are available when required for extra protection against possible cell damage from hailstones, dust, salt air, and similar environmental influences.

Another important consideration in the design of a solar panel is the leads that make the electrical connections to the cells. Normally, the metal used to make these electrical connections would be subject to corrosion. In a well designed panel, such as the one shown in Fig. 4-8, the metallization used to make the electric connections in the panel is completely protected from the environment by a Teflon coating. This type of panel construction provides a very durable,

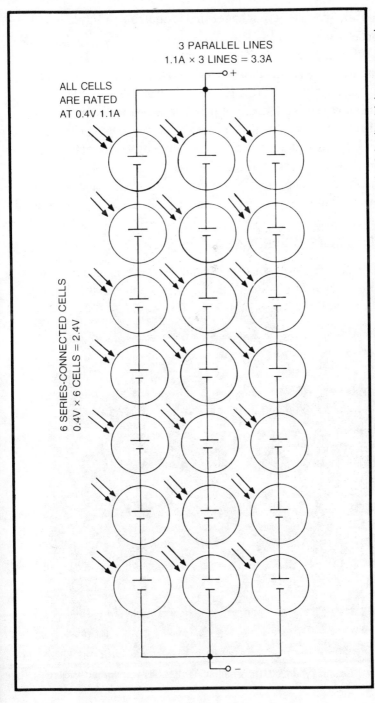

Fig. 4-7. Series-parallel connection of solar cells to provide any voltage and current rating. These cells are rated at 0.4 volts at 1.1 amperes each.

weather-resistant package that requires practically no maintenance.

Looking at the solar panel illustrated in Fig. 4-8 shows that there is one limitation to using a circular cell as an element of an assembly. This is the fact that circular cells placed side by side leave a great amount of panel area that is illuminated by sunlight, but is not used to convert sunlight into electrical power. This limitation can be overcome by using rectangular cells that can be nested together much more closely so that almost all of the panel area is effective in producing electrical power. Panels of this type are described in the following paragraphs.

HIGH DENSITY SOLAR PANELS

Many different solar-cell shapes have been used to make more efficient utilization of the area of a panel that is illumi-

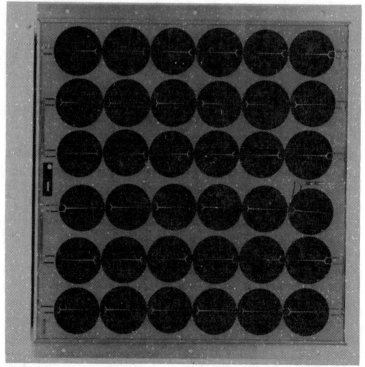

Fig. 4-8. Solarex Type 4200J Unipanel, proides 22 watts at a nominal 14 volts (20 Voc).

Fig. 4-9. Solarex Type 1480 Unipanel uses semicircular cells.

nated by sunlight. Figure 4-9 shows a panel in which the individual cells are semicircular in shape. This permits increasing the amount of active cell area on a panel of given size. This particular panel provides a power output of 10 watts at a nominal terminal voltage of 14 volts under full sun conditions.

Much more efficient use of panel area is possible with cells that are square in shape. Figure 4-10 shows a Solarvoltaic® panel in which essentially the entire surface area of the panel is an active photovoltaic area. Even the aluminum frame

Fig. 4-10. High-efficiency Solarvoltaic® panel provides 35 watts at either 14 or 28 volts.

is designed to minimize the inactive area when several panels are used together in a larger array.

In addition to their novel square shape, the cells in the Solarvoltaic® panel have a unique design. The energy conversion efficiency of the individual cells is in the order of sixteen percent and the overall efficiency is thirteen percent or better. The panel shown in Fig. 4-10 has a power output of 35 watts and will provide either 14 volts or 28 volts, which is selectable by connections at a junction box that is a part of the panel.

Figure 4-11 shows the voltage-current curves of the panel pictured in Fig. 4-10. The curves on the left are the familiar voltage-current curves for various amounts of illumination. The curves at the right are all for full illumination with different cell temperatures. With the amount of sunlight found in the average United States location this panel will provide about 1.06 kilowatt-hour per week.

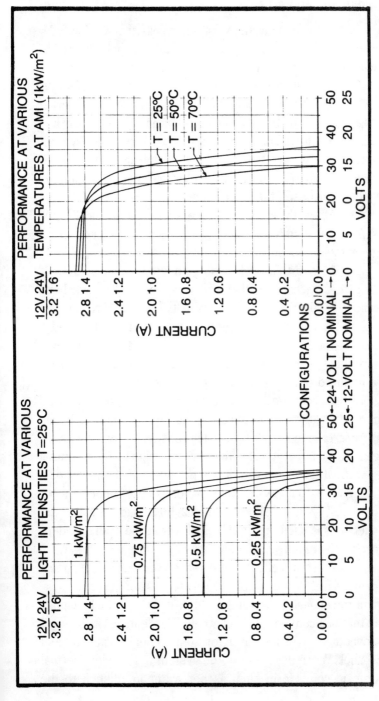

Fig. 4-11. Voltage-current curves for the panel shown in Fig. 4-10.

CONCENTRATOR CELLS

In all of the applications of solar cells that we have discussed thus far, the output of a solar cell or panel is limited by the amount of sunlight that will be intercepted by its area. The question naturally arises as to why we can't use some sort of focusing arrangement that will capture sunlight from a larger area and focus it on the solar cell. The principle is shown in Fig. 4-12. Here a focusing lens, very much like an ordinary magnifiying glass, is used to focus sunligth from an area six inches in diameter onto the surface of a solar cell having a diameter of only three inches. If the power density from the sunlight is 100 mW/cm^2, then the density at the solar cell should be 200 mW/cm^2. Obviously such an arrangement would have advantages in some applications.

The reason that such arrangements haven't been used until rather recently is that ordinary solar cells have been unable to handle the high power density that would result from focusing. A comparatively new type of solar cell, called a concentrator cell, will allow the use of focusing arrangements where the power density at the surface of the cell may be up to two hundred times the normal power density of sunlight.

A typical concentrator cell may have an area of 4 cm^2 to 25 cm^2 or even larger. The thickness usually ranges from 2 mils (50.8 microns) to 14 mils (355.6 microns). The construction of the cell is such that it will withstand very high light concentration without damage and will still provide a linear power output. Figure 4-13 shows the pattern of electric connections to such a cell. Note that the metallization used to make the connections is such that it will block very little of the incident sunlight, thus improving the efficiency of the cell.

Concentrator cells are often assembled into large panels to provide large amounts of power. Because of the concentration of the sunlight, the temperature at the cell will be higher than that of a conventional solar cell. This feature can be used to advantage by using water, or some other coolant, to cool the panel. When this is done, the system will provide thermal energy in some form such as hot water in addition to the

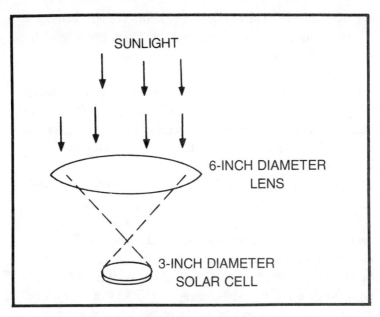

Fig. 4-12. Principal of focusing to concentrate light on a solar cell.

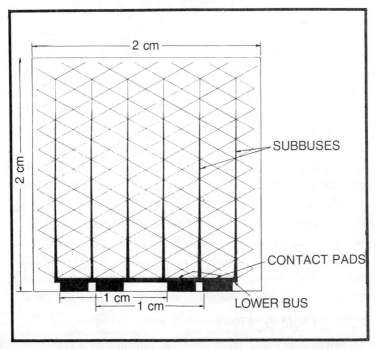

Fig. 4-13. Electric connection pattern of a high-efficiency concentrator cell.

electrical energy. The overall energy conversion efficiency of a concentrator panel can be made very high because of the additional thermal energy.

Figure 4-14 shows the voltage-current curves for a typical concentrator cell under full sun conditions. The various curves are for different cell temperatures. Note that the power output of a concentrator cell is lower at higher cell temperatures, just as in an ordinary solar cell. This shows that if we use a coolant to collect thermal energy in addition to the electrical energy and thus lower the temperature of the cell, we will actually increase the efficiency of the cell.

The efficiency of a concentrator cell also increases with light concentration. This is shown in the graph of Fig. 4-15. Here the concentration ratio is simply the amount of magnification of the incident sunlight. For example, if the concentration is forty times, the power density at surface of the cells will be forty times the power density of the incident sunlight.

Figure 4-16 shows a low concentration array that will provide over 100 watts of peak power. Each of the six panels in the array contains seventeen high efficiency concentration cells with integral mirrored reflectors that provide a geometric concentration ratio of two times. The electrical performance of the array is tabulated below.

Peak power ....112 watts
Voltage at peak power ....48 volts
Current at peak power ....2.3 amperes
Open circuit voltage ....57 volts
Short circuit current ....2.5 amperes

Measurements on the array shown in Fig. 4-16 show that during the daylight hours of a typical day the effective concentration ratio averages over 1.6 times. The energy produced is about 480 watt-hours per day. By way of comparison, a typical flat panel with the same active photovoltaic area would produce only about 284 watt-hours under the same conditions.

SOLAR MICROGENERATORS

We noted earlier that solar cells come in a wide variety of sizes. In addition to the larger units that are used to supply

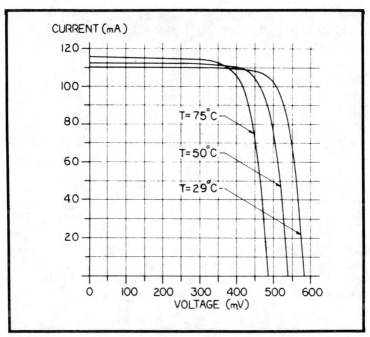

Fig. 4-14. Voltage-current curves for a typical concentrator cell.

large amounts of power, there is a need for very small photovoltaic devices to supply operating power for devices such as electronic watches, calculators, and flashlights.

Figure 4-17 shows an assortment of Microgenerators which are used for this purpose. Most frequently, Mic-

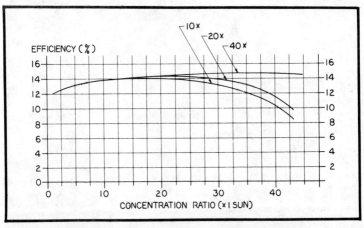

Fig. 4-15. Efficiency of a concentrator cell at various concentration ratios.

Fig. 4-16. Low concentration array.

Fig. 4-17. Assorted Microgenerators used to power calculators and watches.

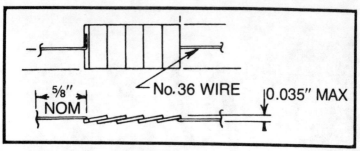

Fig. 4-18. Construction of a Microgenerator array.

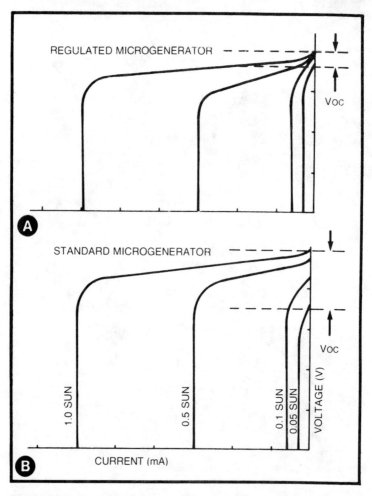

Fig. 4-19. Voltage-current curves of (A) a regulated, and (B) a standard Microgenerator array.

Fig. 4-20. The Solarex three-inch cell and quadrant generate electricity.

rogenerators are used to recharge very small batteries that supply power to devices that use very little power, but there are some devices where all of the operating power comes directly from a Microgenerator.

The assemblies are designed to be very efficient at both high and low light levels. The low light level efficiency is very important in such applications as electronic watches or calculators because these devices are usually used under conditions of low illumination such as shade or room light.

Because of their attractive appearance, Microgenerators are often used as a part of the decorative design of devices such as a watch or a calculator. The silicon surface may have either a shiny or a mat appearance which can be blended in with the overall appearance of the device in which it is used. Microgenerators can also be supplied in colors.

The Microgenerator is actually made up of several extremely small solar cells connected in series. Figure 4-18 shows how the cells are put together into an array. Figure 4-19 shows the voltage-current curves of both a standard Microgenerator and one that includes voltage regulation. These curves will be confusing at first, because the coordinates are different than those that we have been using for solar cells. In this illustration we have plotted voltage along the vertical axis and current along the horizontal axis. Note that with the regulated arrangement the open circuit voltage changes very little with varying light conditions.

Chapter 5
How To Use Solar Cells

Solar cells are very easy to use as compared to most other energy sources. A solar electric system has no moving parts and usually requires little if any maintenance. Some analysis is required in the design of a system, but the process is very simple. A good design can be obtained using only graphs and simple arithmetic. Computer programs are available that will optimize the design of any system, regardless of size.

There are two principal considerations in the design of any solar electric power system. First we have to know how much sunlight is available at the proposed site and how it varies with the seasons of the year. This will tell us the size of the solar electric generator needed to supply any given amount of power. Next, we must know the characteristics of the load including the average current requirement and the duty cycle. This tells us how much storage battery capacity we will need to keep the system operating when sunlight isn't available.

AVAILABLE SUNLIGHT

Finding the amount of sunlight that reaches the earth in any given location usually doesn't require any measurements.

Government agencies interested in agriculture and weather forecasting have been recording the amount of sunlight in all parts of the country for many years. Usually these records can be used to determine the number of solar cells required for any particular application.

Some caution must be used in applying these figures because they may not apply directly to a solar electric installation. The first problem is that sunlight is usually recorded in terms of a unit called the langley. One langley is equal to 11.62 watt-hours per square meter. It is easy enough to convert from langleys to watt-hours, but there is another consideration. The published figures usually represent the sunlight reaching a square meter of the earth's surface, that is, the energy incident on a horizontal plane. Solar cells are usually tilted so that they intercept the maximum amount of energy all through the year. Therefore the inclined solar cell array will usually intercept more energy than an equivalent horizontal surface flat on the earth.

A much more useful measure of solar radiation is one which expresses the average number of *peak sun hours per day* for a particular location. Figure 5-1 shows such a map. The radiation figures have been corrected to account for the solar cell panel being tilted at an angle of 45 degrees and oriented due South. This has been found to be the optimum orientation for panels in most parts of the United States.

The numbers in Fig. 5-1 are a yearly average. These will be somewhat lower during periods when sunlight is diminished. Figure 5-2 shows a map where the figures are the peak sun hours per day averaged over a *four-week period* from December 7th through January 4th. Note that these figures are somewhat less then those of Fig. 5-1.

In both Figs. 5-1 and 5-2 it has been assumed that the solar cells would not be in the shade during part of the day. This is the preferred way to install a system. There will be some places, however, where it isn't practical to find a spot to mount the solar array where it won't experience some shade, at least part of the time. In such applications allowance must be

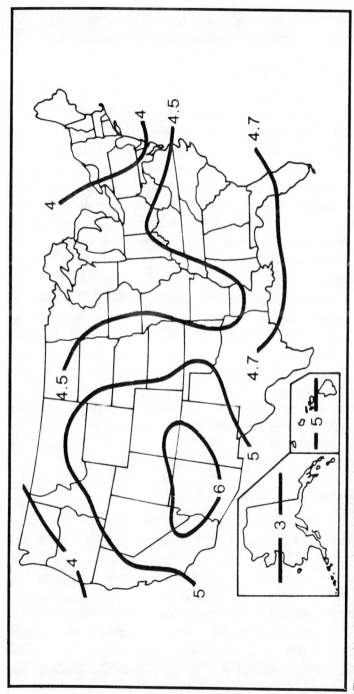

Fig. 5-1. Yearly average peak sun hours per day.

made for the shade. This usually means that more solar cells will be required to furnish a given amount of power.

LOAD REQUIREMENTS

Normally the power rating of an electric device is given in watts. This, of course, is the rate at which the device consumes electrical energy. The total daily energy requirement of a device is specified in watt-hours or kilowatt-hours.

In an earlier chapter we saw that the voltage from a solar electric generator is nearly constant. Most systems include storage batteries which also tend to have a constant voltage output. For this reason, we can quite conveniently specify the power consumption of an electric device in terms of amperes. In this case the daily energy requirement will be in ampere-hours.

By stating the energy requirement in ampere-hours we can easily match a load to a solar generator and determine the size of the solar generator required for that particular load.

If the load is something that draws a constant current, the problem is rather simple. Suppose, for example, that we wish to use solar energy to operate an electric clock that draws a current of 20 milliamperes. The load current in amperes is simply 0.02 amperes and the daily (24-hour) energy requirement in ampere-hours is

$$\text{Ampere-hours} = 0.02 \times 24 = 0.48$$

INTERMITTENT LOADS

Most of the loads that we might wish to operate from a solar electric generator will not draw a constant current. For example, a two-way radio station draws much more current when it is transmitting than when it is receiving. Similarly, a water pump usually pumps on demand, drawing maximum current while it is pumping and practically no current while it is idling. Of course, the varying voltage and current requirements of different applications are no problem because solar cells can be connected in any series-parallel configuration required to provide the necessary voltage and current.

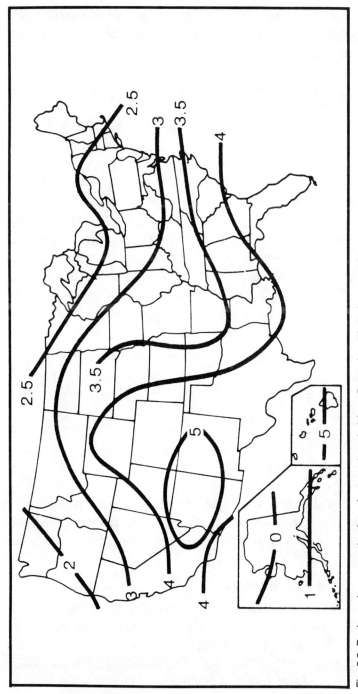

Fig. 5-2. Peak sun hours per day for a four-week period from December 7th to January 4th.

The easiest way to determine the average current, as well as the total daily load in ampere-hours, is to make a tabulation of the amount of current drawn at various times over a 24-hour period. Once this information is available, the design of the system is quite simple as shown in the following example.

We have a two-way radio repeater which is located on a remote mountain top where its antenna will provide good coverage. The area is not near any power lines so we wish to power the repeater with a solar electric generator. When the repeater is transmitting, it draws 5 amperes of current. While it is idling, it only draws about 1 ampere. A study of the system shows that it only transmits about ten percent of the time during any 24-hour period. Knowing these facts we can find the number of ampere-hours required for a 24-hour period.

$$\text{Transmitting } 5 \times 0.10 \times 24 = 12$$
$$\text{Idling } 1 \times 0.90 \times 24 = 21.6$$
$$\text{Total } 12 + 21.6 = 33.6$$

Thus, with very little effort we have found that the average daily load of the repeater is 33.6 ampere-hours. To find the average current, we merely divide this figure by twenty-four hours.

$$\text{Average amperes} = \frac{33.6}{24} = 1.4$$

A tabulation of this type can be prepared for any type of load. Once we have the load requirement, we can determine the size of solar generator required to do this job.

SYSTEM CALCULATIONS

Knowing the average number of peak sun hours at a proposed site and the average daily load requirements in ampere-hours, we can quite readily calculate the total current output necessary from the solar generator. The following equation shows how this is done:

$$\frac{\text{Average daily load}}{\text{Peak sun hours}} \times 1.2 = \text{Amperes required}$$

The factor of 1.2 is included to take care of inevitable system losses. The number of solar panels needed to provide this current can be found by dividing the total current by the current output of an individual panel. That is,

$$\text{Total panels} = \frac{\text{Amperes required}}{\text{One-panel current}}$$

To illustrate the use of these equations, consider a two-way radio repeater system installed in Kansas City, Missouri. The repeater operates continuously, twenty-four hours a day, and draws an average current of 2 amperes. Reference to Fig. 5-1 shows a daily average of 4.5 peak sun hours at the site. Using the first of the above equations, we find

$$\text{Amperes required} = \frac{2 \times 24}{4.5} \times 1.2 = 12.8$$

The nominal operating voltage of the repeater is 12.8 volts so the solar panels will be used to maintain a charge in a 12-volt battery. It has been found through experience that a solar generator with a nominal terminal voltage of 14 volts is well suited to such an application. Thus the Solarex Type 4200 Unipanel, which has an output current of 1.3 amperes at a nominal terminal voltage of 14 volts, was selected as the basic element of the solar array. Using the second equation given above, we find that

$$\text{Total panels} = \frac{12.8}{1.3} = 9.8$$

In practice the number 9.8 is rounded off to ten, and the generator for the system consists of ten Type 4200 Unipanels connected in parallel.

Note that we selected a panel that produced the required voltage and used our equations to find the number of panels required for the application.

If the repeater drew the same current as the one in our example, but operated at 24 volts, we use ten more Unipanels in series with those in the example. This would give us a nominal terminal voltage of 28 volts, which has been found to be suitable for use with 24-volt batteries.

This type of system calculation will lead to a conservative design that will be highly satisfactory in most applications. It is possible to make a much more detailed analysis by using the equations to predict the performance for every month of the year. In some instances this may result in a more economical design. A computer program for the purpose is available from Solarex.

BATTERIES

In the preceding discussion of load requirements we considered the load requirement over a 24-hour period without regard to the fact that sunlight will not be available during the full 24-hour day. Any solar generator which might be called on to furnish energy at a time when light is not available must include a storage battery.

The function of a battery in a solar electric generator is twofold. The battery must furnish power to the load when sunlight is not available, and it also must smooth out variations in the load. All the solar panel has to do is to furnish the average daily load requirement. Naturally, all of this energy will be generated during daylight hours. The battery will store the energy and make it available to the load as required.

Many different types of batteries are available, each having its advantages and limitations. Batteries are usually classified into two general types—primary batteries and secondary batteries. The primary battery is one that can not be recharged and is of no interest in solar electric generators. The secondary battery is a storage battery that can be recharged many times and is the type in which we are interested. Strictly speaking, the word battery should apply to a collection of separate cells, but common usage applies the term to any device that furnishes electrical energy, even though it might be a single cell. The ordinary flashlight battery is a good example. It is almost always called a battery, although it is really a single cell.

There are many different types of storage batteries commercially available. Selection of a battery type for a particular solar electric generator involves many considerations. Included among these are:

- Voltage requirement
- Current requirement
- Operating schedule
- Ampere-hour capacity
- Operating temperature range
- Size and weight
- Required life
- Cost

TYPES OF BATTERIES

By far the most common type of battery used in solar electric generators is the lead-acid battery. The familiar automobile battery is a good example of this type. Until comparatively recently, the lead-acid battery was a vented unit with removable caps used to refill the electrolyte. This type of battery is probably still the most economical type. It can deliver relatively large currents at a nearly constant voltage for long periods of time. It has the limitations that it requires periodic maintenance, operates poorly at very low temperatures, and vents hydrogen gas during the charging period. In some applications, the hydrogen gas may constitute an explosion hazard.

Several different types of sealed lead acid batteries have become available in recent years. These are usually more expensive than the vented battery, but the saving in maintenance can often justify the additional cost. Sealed automobile batteries carrying a five-year maintenance-free warranty are readily available. These batteries are well suited for use in many solar electric generators.

One type of sealed lead-acid battery uses a gelled electrolyte. Although this battery is sealed, the seal is a resealing type that will vent gasses if the internal pressure should become too great. Thus there is no danger of explosion if something should go wrong in the system. During the normal charge and discharge cycles, the pressure never becomes great enough to operate the vent.

There are two separate types of batteries using gelled electrolytes. One type is designed for the battery to be floated across a line at all times so that it will furnish power only when the regular power source fails. The other type is designed for regular charge and discharge cycles.

There is another type of sealed lead-acid battery called a "starved electrolyte" type. Again this type of a battery is vented with a resealable vent, but usually the pressure is much higher than in the type described above. As a result, this type will withstand higher current discharges without loss of electrolyte.

Still another type of sealed lead-acid battery is truly sealed. The chemistry of the battery is such that under conditions that would normally increase the internal pressure in the cell, the process is stopped before the internal pressure can become excessive.

For smaller loads, the most common type of storage battery used is the sealed nickel-cadmium battery. This is the familiar rechargeable battery that is used in calculators and portable walkie-talkie radios. One of the disadvantages of the nickel cadmium, or nicad as it is often called, is that it tends to develop a memory. If it is continually operated with the same charge-discharge pattern, it will want to follow this same pattern. For example, if a nicad is operated in an application where it is allowed to discharge about twenty-five percent before recharging, it will tend to loose its capacity to be discharged lower than about twenty-five percent.

Two other types of storage battery—silver cadmium and silver zinc—have a very high energy density; that is, they will produce more kilowatt-hours of energy per pound of battery than other types. They are more expensive but well suited to applications where a very small battery is required and it is desirable to get as much energy as possible stored in the small volume. Typical applications for these batteries include portable electronic equipment and military devices. The silver-zinc battery, sometimes called a silver-oxide battery is used frequently in electronic watches.

BATTERY SPECIFICATIONS

Earlier we listed some of the factors that must be considered in selecting a battery for any given solar electric generator. In practice, the best policy is to contact the battery manufacturer for a recommended type for any application. The following information will give a general idea of how the various requirements of an application affect the choice of a type of battery.

Voltage Requirement

It goes without saying that a battery must not supply voltages high enough to damage the equipment that is being powered. This is usually prevented by the use of voltage regulating circuits in the equipment. Another voltage consideration centers around the fact that many electronic items will not operate at all when the supply voltage falls below a certain value. Figure 5-3 shows how the voltage of nickel-cadmium batteries falls off as the battery becomes discharged. Note that the terminal voltage of the nicad holds comparatively constant through much of the discharge cycle. This shows that the nicad might be the better choice in applications where the low end voltage of the cycle is critical.

Current Requirement

In some applications the current drain will be nearly constant. In others the system will be required to supply a large current for a short period of time. Usually the sealed nicad is best suited for supplying short bursts of high current, but the lead-acid battery will also handle such requirements.

Operating Temperature

Most storage batteries will work well at reasonably low temperatures. The lead-acid battery will operate at lower temperatures than the nicad. In either case, the output of the battery tends to be lower at lower temperatures. For this reason, any application where the batteries will be exposed to low temperatures, as at high latitudes, will require additional battery capacity because of this factor.

Size & Weight

In many applications of solar electric generators there is plenty of room for the batteries so this is not a problem. A notable exception is when a Microgenerator is used to power an electronic watch. Here space is at an absolute premium and special batteries fabricated for the application must be used. The silver-zinc battery is widely used for this purpose.

Ampere-Hour Capacity

Probably the most important consideration in the selection of a battery for a solar electric generator is the ampere-hour capacity. Simply stated, the battery must have enough ampere-hour capacity to power the load until there is enough sunlight to recharge the battery. In a small application, the situation is often simplified because it is easy to make the battery oversized. Then there will always be enough energy in the battery to operate the load.

In a larger installation where the batteries represent a considerable investment, it is usually worthwhile to make a rather detailed analysis of the requirement so that the investment in batteries will not be higher than necessary.

The ampere-hour capacity of a battery is usually specified together with some standard hour reference such as ten or

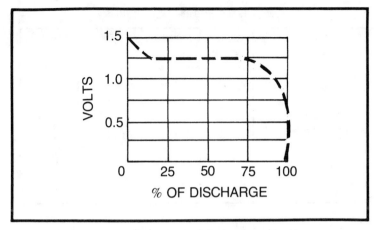

Fig. 5-3. Terminal voltage versus depth of discharge for nicad battery.

twenty hours. What this means is that the battery has a certain ampere-hour capacity if it is discharged at a certain rate. If the current drain is lower, the capacity will be somewhat lower. For example, suppose that a battery is rated at 80 ampere-hours and a ten-hour reference is specified. This means that when the battery is fully charged, it will deliver a current of 8 amperes for ten hours. If the current is higher than 8 amperes, the capacity will be somewhat lower. The relationship between the capacity of a battery and the load current can be found in the manufacturer's literature.

The battery field is not static. New types are being developed regularly for different applications. For this reason, no list of specifications and features of various batteries can be complete. A list of some of the more commonly used batteries and their characteristics is given in Table 5-1.

THE COMPLETE SYSTEM

Figure 5-4 shows a diagram of a complete solar electric generator. Note that a diode has been included between the solar cells and the storage battery. The purpose of the diode is to assure that current will only flow when the solar cell is delivering energy. If the diode were not included, whenever the voltage produced by the solar cell was zero, battery cur-

Table 5-1. Storage Battery Characteristics

TYPE OF BATTERY	VOLTS PER CELL	NOMINAL CAPACITY (A-H)	DEEP DISCHARGE CYCLES	EST LIFE (YEARS)
NICad, SEALED SINTERED PLATES	1.2	0.2–160	200–2000	2–10
NICad, SEALED POCKET PLATES	1.2	0.05–70	100–250	5
LEAD ACID SEALED	2.0	0.9–7.5	100–400	4–5
LEAD ACID VENTED	2.0	0.1–120	200–700	3–6
SILVER-ZINC	1.5	0.1–300	1–80	0.02–2

rent would flow through the solar cell. This would discharge the battery and might damage the solar cell.

In applications where the solar panel will deliver more energy than is required by the load at any particular time, it is advisable to include a regulator to prevent overcharging of the batteries.

As an example of selecting a battery for a system, suppose that the device that we wished to operate with solar electricity required 12 volts and had an average daily requirement of 10 ampere-hours. Further suppose that the longest period where there would be little sunlight is expected to be

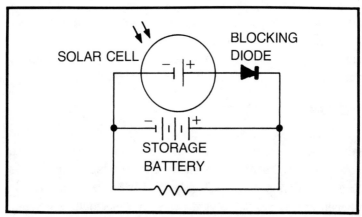

Fig. 5-4. Complete solar electric generator. The diode prevents battery current from flowing through the solar cells when its output is zero.

Fig. 5-5. An array of solar panels for receiving sunlight and generating electricity.

one week. To be on the safe side, the batteries should be capable of operating the sytem for a week. We could meet this requirement quite easily by using an 80-ampere-hour automobile battery. If the system were located where the average temperature during part of the year would be very low, it would be advisable to include additional battery capacity.

Chapter 6
Applications of Solar Cells

Originally the solar electric generator, being comparatively expensive, was used only where other sources of electrical power were not practical. Thus the first significant application of solar cells was in the space program. In fact much of the solar cell development performed under the space program is finding its way into other applications.

Recently, the picture has changed drastically. Whereas the cost of conventional hydrocarbon fuels is rising sharply, the cost of solar electric generators is actually declining. This, together with the threatened shortage of conventional fuels, is leading to many new applications of solar electric generators. Solar cells are now used, not only in locations where power lines are not available, but also in many places where they are simply more convenient and reliable. In several applications the solar electric generator is becoming the most economical source of electric power.

In this chapter we will discuss a few present day applications of solar cells. The examples have been selected to give an idea of the wide variety of applications of solar cells. The descriptions given here by no means exhaust the possible applications, which seem to be limited only by the ingenuity of the user.

RADIO & TELEVISION

The transmission of radio and television signals depends on the use of an optimum site for the transmitting antenna. The best site is usually the one at the highest possible elevation. Unfortunately, such a site is apt to be one where electric power lines are not available and the transportation of fuel for portable electric generators would be prohibitively expensive.

Much of the two-way radio in use today depends on the use of radio repeaters. A repeater is simply a receiver and transmitter located at some high point overlooking the area of signal coverage. Radio transmitters installed in motor vehicles use small antennas and are often not in good locations. As a result their signals will not propagate over large distances. By using a repeater at a nearly optimum site, the weak signals from the mobile units can be picked up and rebroadcast to provide good coverage over a large area.

In the past, some of the best possible sites for repeaters have been passed up simply because electrical power was not available at the site. In many areas there are hills or mountains that would be ideal repeater sites, but because of the nature of the terrain power lines have never been brought to the area.

By using a solar electric generator almost any accessible site can be used as a repeater location. Figure 6-1 shows a system installed in the Mojave desert to power both microwave links and two-way VHF radio repeaters used by the California Highway Patrol. The solar array has a peak power rating of 460 watts. Because of the fact that the solar cells will operate reliably at very high temperatures, the array is almost maintenance free.

An FM or TV translator is very similar to a two-way radio repeater, except that it rebroadcasts radio and TV programs to areas which, because of their geographical location, would otherwise have little or no radio or TV service. The translator usually picks up a signal on one channel and rebroadcasts it on another channel that is not used for broadcasting in the particular area.

Fig. 6-1. Solar-powered radio repeater on the Mojave desert in California.

As was the case with repeaters, many ideal sites for translators have never been used because of the absence of electrical power lines. Sites have usually been chosen where power lines were available even if the site happened to be far from ideal for reception or transmission of signals. Solar electric generators make it practical to select ideal sites regardless of the availability of power lines.

Solar energy can also be used for powering radio and TV receivers. More and more public service announcements of hazardous weather conditions and other emergencies are being made on television broadcast stations. These announcements are of little value if the people affected are without electrical power during the emergency. A solar electric generator can be used to operate a radio or TV set both to conserve conventional sources of energy and to provide reliable operation during a power outage. Figure 6-2 shows an experimental arrangement where solar cells are being used to provide operating power for a TV set.

Yet another application where solar cells can improve television reception is in providing operating power for the head end of a cable TV system. The head end of the system is where the signals from distant television stations are picked up for rebroadcast along the cable. As with repeaters and translators, the geographical location of the head end of a cable TV system is very important. Ideally, the head end should be located at the highest practical elevation where the path to the distant stations will be free of obstructions. In the past, the availability of electric power was a prime consideration in the location of the head end. This is no longer a restriction, because solar electric generators can be used to supply the operating power.

FIBER OPTICS

Optical fibers are competing increasingly with coaxial cables for the transmission of data, television programs, and telephone conversations. Both fiber optic systems and the conventional coaxial cable require repeater amplifiers at inter-

Fig. 6-2. Experimental solar panel operates TV set.

vals along their length. The coaxial cable has the advantage that operating power for the repeater amplifier can be transmitted along the cable with the signal. The optical fiber, being an insulator, can't get power to the repeater in this way. Thus a power source is required at each repeater amplifier location. This need not be a problem, because small solar electric generators can be used at each repeater location. Figure 6-3 shows a sketch of a fiber optic repeater with a solar electric generator.

AGRICULTURAL APPLICATIONS

There are many places on agricultural installations where solar electric power can be used to advantage. In recent years many different electric and electronic devices have been developed that are revolutionizing all aspects of agriculture. The application of these devices is somewhat limited by the fact that a farm occupies a large area and electrical power is not available over much of the area. Solar electricity makes it possible to use electric and electronic devices at any location where sunlight is available.

A problem frequently encountered in agriculture is that in many areas of the world where there is an abundance of sunshine there is little or no water for irrigation. Crops require both sunshine and water for growth. Solar electricity permits using some of the sunshine that is plentiful to develop electric power for pumping water which is scarce.

Figure 6-4 shows the world's largest application of a solar electric generator to the problem of irrigating an agricultural installation. This system, installed at the U.S. Department of Agriculture's Mead Field Laboratory in Nebraska, pumps water at the rate of one thousand gallons per minute from an irrigation reservoir for twelve hours every day.

The pumping system, which is driven by a ten horsepower motor, operates on electric power produced by an array of 120,000 individual solar cells. The cells are in the form of Type 9200J Unipanels. This panel is very similar to the Type 4200J which was described in an earlier chapter. The array has a peak power rating of 25 killowatts.

Water pumped by the system is distributed throughout the area by means of a system of pipes, part of which is shown in Fig. 6-5.

Farms use many machines, such as tractors and harvesters, that use conventional fuels such as gasoline to power their engines. These machines also use electric batteries to start the engines and power accessories. To conserve fuel, these machines are often left at remote locations, far from a source of electrical power. If the machine isn't used for an extended period of time because of some such factor such as inclement weather, the storage battery will often go dead making it impossible to start the machine.

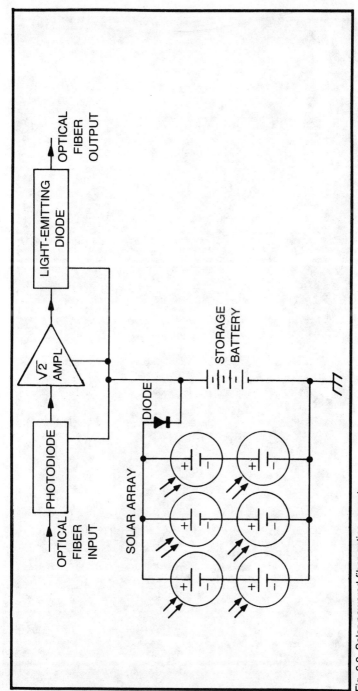

Fig. 6-3. Solar powered fiber optics repeater.

Fig. 6-4. Solar powered irrigation system in Nebraska.

By installing a small solar electric generator on such a machine, the storage battery can be kept fully charged. The machine can then be started easily, even if it has been left idle for an extended period.

Many farms are located in remote areas where electrical power is not as reliable as in an urban area. Power outages are not infrequent. A solar electric generator will provide a good source of emergency power in such locations. By using solar cells to charge storage batteries, energy will be stored during periods of sunshine for use when power is not available during a storm.

THE CONSTRUCTION INDUSTRY

In many areas, construction of new buildings is started long before electrical power lines are installed to the site. There is the same problem of keeping construction vehicles' batteries charged as was mentioned in connection with agriculture in the preceding paragraphs.

Fig. 6-5. Part of water distribution system of Fig. 6-4.

This absence of electrical power can seriously hamper construction activities. Often temporary power lines are strung to the site at great expense.

A temporary solar electric generator installed at a construction site can be used to proivide electrical power during the construction phase of the project. When the job is completed the generator can be moved to the next site.

A common use of solar electricity in construction is the solar powered roadside construction flasher shown in Fig. 6-6.

This device consists of a flashing light, a storage battery, and a small array of solar cells. The cells perform the double duty of keeping the storage battery charged, and turning on the flasher when the ambient light falls below a certain level.

REMOTE AREAS

In most urban and suburban areas, electricity is taken for granted. One of the things that has held up the development and utilization of many areas is simply the absence of electrical power. Of course, portable generators driven by hydrocarbon fuels could be, and somtimes are, used at such locations, but there is still the problem of transporting fuel to the site.

There are many places where it is obvious that electric and electronic devices could be used to advantage if electrical power were available. There are other situations, however, where the absence of electrical power has been accepted for so long that the advantages of electrical solutions to problems hasn't even been considered.

A typical example of this type is shown in Fig. 6-7. Here a solar electric generator is used to power a specially designed flush toilet at a remote site in Custer National Forest. The site is at Rocky Creek Vista Point, a scenic overlook on the Beartooth Highway between Red Lodge, Montana and Yellowstone National Park. It is estimated that nearly 50,000 people visit this spot during the summer months. Until recently, the only sanitary facility at the site was an eight-fixture vault toilet with an 18,000 gallon holding tank. The vault

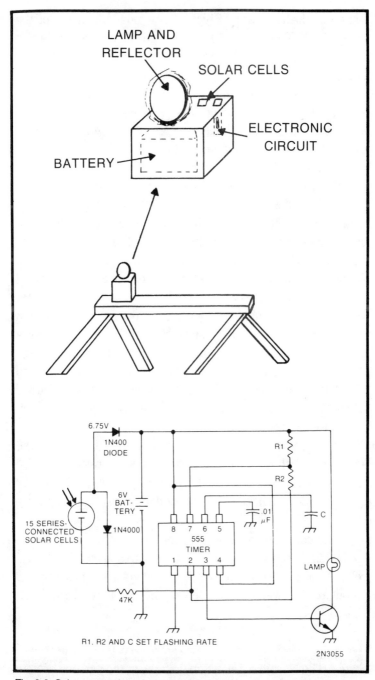

Fig. 6-6. Solar powered road construction flasher.

leaked at the rear wall, which was exposed because of the steep slope. The exposed raw sewage was a potential health hazard, and the strong odors detracted esthetically from the site.

Obviously, a different type of sanitation system was desirable, but neither water nor electrical power were available at the site. The cost of running an electrical power line through the rugged terrain would be prohibitive.

A new type of sanitation system that doesn't require water became available. This system operates much like a conventional flush toilet, but uses a colorless liquid similar to mineral oil for flushing. The fluid looks very much like water, but it can be recycled within the system to carry the wastes to a sealed tank. This flush system uses a small pump, driven by a 24-volt DC motor at each fixture to recirculate the fluid.

Inasmuch as there was no electrical power at the site, various ways of supplying operating power for the system were considered. The closest electric power line was nearly two miles away over formidable terrain. Running a power line to the site would cost somewhere between $35,000 and $85,000. Other power sources that were considered included a gasoline generator with automatic starting, a wind driven generator, and solar electricity.

The storage battery requirements turned out to be the same for all three of these approaches. A rather detailed analysis was made and solar electric panels were finally selected because of the low maintenance cost, and simplicity of operation.

The complete solar array, mounted on the roof of one of the buildings shown in Fig. 6-7, contains 1664 individual solar cells. It produces 15 amperes at 24 volts—a peak power output of 360 watts. The area of the complete array is 73 square feet.

Another example of the application of solar electricity to a remote area is in the largest Indian reservation in the United States. The Navajo reservation, located at the junction of New Mexico, Arizona, Utah, and Colorado, has huts and frame

Fig. 6-7. Solar powered sanitation system in Custer National Forest.

houses scattered over an area that is larger than many eastern states.

Until recently, water was hauled from a community well to the individual homes, and the only form of artificial light was the kerosene lamp. Faced with the problem of improving sanitation and living conditions without the enormous cost of running power lines to the area, the Indian Health Service sponsored the development of solar electric generators for the individual homes.

The system designed for each home, shown in Fig. 6-8, consisted of a Type 4200J Unipanel, having a power output of 20 watts, a 15-watt fluorescent lamp, a three-gallon-per-minute water pump, and an 80-ampere-hour storage battery. Each home installation cost about $400, far less than the cost of running a power line. Several systems have been installed and have proven to be successful.

In the underdeveloped nations of the world, the problem of using electrical and electronic devices is much more severe than in the United States. Although there are many places in the United States where it would be expensive to make power lines available, in some nations there are places where the nearest power line is hundreds of miles away. In such areas, solar electric generators are just about the only answer to providing electrical energy. Figure 6-9 shows a solar powered water pumping system that has been installed in Mali, West Africa. This system is used for pumping water in a village and uses a 600-peak-watt solar panel.

SOLAR ELECTRIC CATHODIC PROTECTION

Electrolytic corrosion is one of the major causes of deterioration of all types of metal structures. It has been estimated that the annual cost of corrosion in the United States is about eight billion dollars.

Electrolytic corrosion can occur whenever we have two dissimilar metals in contact with each other, or connected together electrically, under conditions where the environment may be electrolytic. As shown in Fig. 6-10, one metal will tend

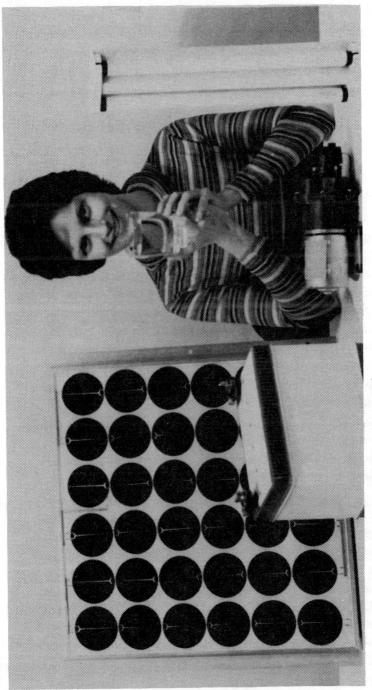

Fig. 6-8. Lighting and water pumping system for Navajo reservation.

to become an anode, and the other a cathode, of what amounts to a small electric battery. Whenever a current can flow from the cathode of this battery to its anode, either through the environment or along the surfaces of the metals, corrosion will occur at the anode. Note that there is no corrosion at the cathode.

Actually, we don't need to have two different types of metal to get electrolytic corrosion. Sometimes, impurities in the same type of metal can make one part behave like an anode and another part like a cathode. The electrolytic currents can eat away a large amount of metal at the anode. For example, a current of 1 ampere will corrode away twenty-two pounds of steel in a year.

One way to protect against electrolytic corrosion is to force an appropriate amount of current to flow through the metal structure in the opposite direction to which corrosion currents would flow. This will make the part of the metal that would otherwise behave like an anode to be a cathode, at which no corrosion will take place. Another metal, which we will allow to corrode, can be added to the system to serve as an anode. This other type of metal that is added to the system is often called a sacrificial electrode because we will allow it to corrode instead of the main structure. This method of protection against electrolytic corrosion is usually called *cathodic protection*. The method requires a source of electrical energy which can quite conveniently be a solar electric generator.

One of the first applications of solar electricity to cathodic protection was made under the sponsorship of the U. S. Department of Transportation on a bridge near Washington, D.C. in 1976. This system is in use on the Dear Run Bridge on the George Washington Memorial Parkway. Solar electricity was selected as a power source because the initial cost was $4,620 as compared to the proposed cost of $13,954 for using commercial power. In addition, with the solar system there is no monthly charge for electricity.

An analysis of the structure showed that a continuous direct current of 6 amperes at 2 volts would be required to

Fig. 6-9. Solar array provides power for pumping water in Mali, West Africa.

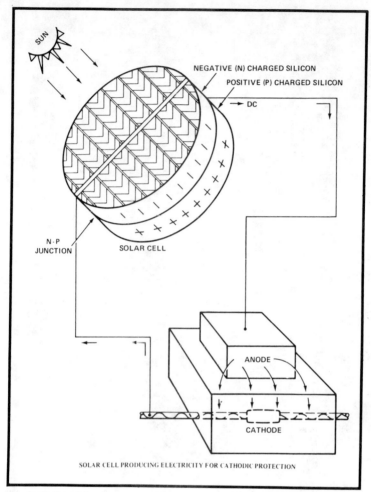

Fig. 6-10. Principle of electrolytic corrosion.

provide cathodic protection. A study of the weather data in the Washington, D. C. area showed that in order to get a continuous 6-ampere current twenty-four hours a day, a peak current of 33 amperes would be required. Figure 6-11 shows a sketch of the system. Energy from the solar array charges a bank of batteries which in turn furnish current to the electrodes of the protection system. The usual blocking diode is provided to prevent the batteries from discharging through the solar cells during the night. In this application, no charge limiting system was required. The amount of sunlight was not great enough to overcharge the batteries during periods of sunlight.

The storage battery selected has a capacity of 2016 ampere-hours. It is expected that the only maintenance required by the system will be adding water to the battery once every two or three years.

The solar electric generator selected for this application consists of ten panels connected to provide the required 33 amperes of peak current. The nominal voltage output of the array is 2.4 volts, which is high enough to allow efficient charging of the 2-volt battery.

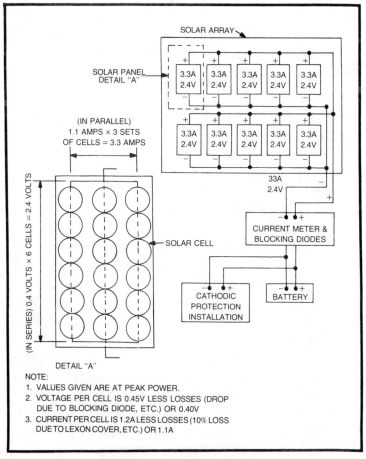

Fig. 6-11. Solar powered cathodic protection system.

APPLICATIONS OF CONCENTRATOR CELLS

The concentrator cell described in Chapter 4 not only reduces the number of cells required for a given power output, but also increases the efficiency of the array and reduces the cost per watt of electrical power. The fact that thermal energy can also be derived from the array further increases the overall energy conversion efficiency of the system.

One of the largest applications of concentrator cells is at Mississippi County Community College (Fig. 6-12) in Blytheville, Arkansas. This installation, funded by the U. S. Department of Energy, contains more than 50,000 concentrator cells encapsulated in nearly 300 solar modules. The system provides approximately 362 kilowatts of peak power and supplies energy for new solar powered classroom facilities.

MILITARY APPLICATIONS

The armed forces are the largest users of electronic equipment and are always in need of reliable sources of electrical power. The most obvious need is at battle sites where conventional power lines may not be available or may have been destroyed. The logistics problems of transporting fuel for portable generators to such sites are often severe.

A less obvious need for power is for testing facilities, where tests of armament are necessarily carried out at remote locations. The cost of running power lines to some of these testing sites is prohibitively high. Solar electric generators provide a reliable source of power that can be used to recharge batteries and reduce the need for fuel for powering portable electric generators.

Figure 6-13 shows a solar electric generator installed at Luke Air Force Base in Arizona. This system supplies operating power to a TV camera and its associated equipment used as a part of the Television Optical Scoring System (TOSS). This system is used to evaluate the performance of weapons systems. Because the system is self-contained, it can be mounted very close to the range. No regular operating or

Fig. 6-12. One of the largest installations of solar concentrators located at Mississippi County Community College in Blytheville, Arkansas.

Fig. 6-13. Solar cells power military range instrumentation system.

maintenance attention is required, so there is no need for protective bunkers for operating personnel.

MICROGENERATOR APPLICATIONS

Normally, when we think about the most expensive applications of electric power, we think of installations that use huge amounts of power, such as aluminum refineries. Actually, although such installations use large amounts of energy, the cost per kilowatt-hour is usually very low. The applications where the cost of electrical energy is inconceivably high is with very small devices that use minute amounts of energy. A good example is the electronic watch.

If the battery used in an electronic watch should cost about $1.00, the cost of the actual energy supplied by the battery would be about $4764.00 per kilowatt-hour. This compares with the regular electric utility charge of a few cents per kilowatt-hour of energy.

In spite of the efforts of the designers of batteries for such things as electronic watches, the life of the battery is often less than a year. In an effort to reduce the cost and improve the life of watch power supplies, many manufacturers are using or planning to use solar electric generators in connection with a rechargable battery to power electronic

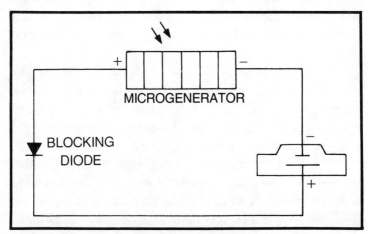

Fig. 6-14. Microgenerator battery charging circuit.

watches. By using a Microgenerator with one of the newer silver-oxide batteries, a life span of five years or greater can be obtained. Figure 6-14 shows a typical circuit.

The small electronic calculator, although usually not as small as an electronic watch, has a similar power supply problem. Advances in electronics have made it possible to reduce the size of the calculator continually. As the calculator becomes smaller, the battery is apt to become the weakest link in the system. If the battery size is continually reduced, the battery capacity and hence the life of the battery will also be reduced. The solar Microgenerator array is an ideal power source for devices of this type. It can be used to recharge a storage battery, or as we will see in the following paragraph, to operate the calculator without any battery at all.

The Microgenerator is designed to have very good energy conversion efficiency at both high and low light levels. As a result of its low light level efficiency, it can provide enough energy to operate a calculator when illuminated by ordinary room light.

Figure 6-15 shows the world's first solar powered electronic calculator. Using low current drain liquid crystal display units, and a high efficiency Microgenerator as a power source, this calculator doesn't have any battery at all. Neither does it have an on-off switch. When the calculator is subject to enough light to enable the user to read the display, the miniature solar electric generator will provide enough energy for its operation.

There is no need at all for an on-off switch—an item that is subject to wear, and hence to failure—because there is no battery to run down. The calculator will be energized whenever light falls on it, but this is no problem. All of the solid-state components have long lives.

There are many other potential applications of the Microgenerator in electronic devices. Many small instruments are being developed for medical applications. Most of these have to be plugged into a charger periodically to keep the batteries charged. The combination of MOS solid-state technology and

Fig. 6-15. Calculator powered by solar cells has no battery.

liquid crystal readout devices make it possible for such devices to have a very low power drain. Thus they are suitable for use with miniature solar electric generators as power sources.

SOLAR POWERED FLASHLIGHT

Lighting devices that are only used at night are ideal candidates for solar electric power. During the daytime when the device isn't being used, its batteries can be recharged by solar electricity. A device of this type is the solar powered flashlight shown in Fig. 6-16.

The solar powered flashlight looks like any other flashlight except that is has a plug in the back for connecting the solar cell. The batteries are rechargeable, rather than the usual flashlight batteries. When the flashlight isn't being used during the daylight hours, the solar cell is plugged in and recharges the batteries. This not only avoids the inconvenient situation where the flashlight batteries go dead, but it eliminates the expense of replacement batteries.

SOLAR POWERED ATTIC FAN

The solar powered attic fan, shown in Fig. 6-17, is an example of the application of solar electricity to a device that usually only operates when the sun is shining. Usually an attic fan is operated when the sun is shining brightly heating the attic. At night, when there is no sunlight, the fan is usually turned off.

The fan shown in Fig. 6-17 is driven by a 4-volt solar panel that furnishes 6 peak watts. It can deliver about 250 cubic feet of air per minute. Inasmuch as it is only operates when the sun is shining, there is no storage battery.

SOLAR POWER FOR RECREATION

Many recreational activities, such as golf, hunting, and boating, involve the use of large areas of land or water. Providing electric power lines to such large areas is always expensive, and sometimes impossible. Here again, solar electricity can come to the rescue and make it practical to use electrical or electronic devices at almost any location.

One of the earliest applications of solar electricity to a recreational activity was in the solar powered electric golf cart. A solar electric generator mounted on the golf cart not only keeps the batteries charged, it eliminates the need to run power lines to any location where the golf carts might be stored when not in use.

Hunting lodges are necessarily located in remote locations close to the game being hunted. The operator of such a lodge is faced with the alternatives of going to great expense to provide electric power, either by means of power lines or portable generators, or letting his guests do without the convenience of electric and electronic devices. This situation can be alleviated by the use of solar electric power.

A solar electric generator mounted on the roof of a cabin can provide operating power for radios, television sets, refrigerators, water pumps, attic fans, and a host of other conveniences and luxuries.

Another example of the application of solar electricity to recreational activities is the solar powered battery charger designed for marine applications. On power boats, an important consideration is conserving on-board fuel. Recharging of

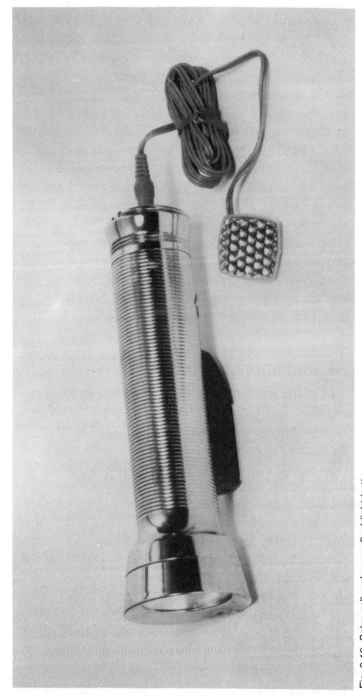

Fig. 6-16. Solar cell recharges flashlight battery.

batteries from the boat's engine consumes fuel. Many sailboats and yachts have little if any auxiliary power, hence no source of electrical power.

There are many electrical and electronic devices that can be used on boats, not only as conveniences, but as safety devices. The use of a solar electric generator makes it possible to operate these devices without spending any of the on-board fuel. Most boating is done during daylight hours when plenty of sunlight is available. This assumes that the batteries will be fully charged when the sun sets. Thus the supply can be used to operate running lights, a bilge pump, two-way radio, and other electronic gear.

The solar Energizer® designed for marine applications weighs only four pounds and provides 0.6 amperes of current in full sunlight. In the average United States location it will develop about 18 ampere-hours each week. The charger is designed to work with the standard 12-volt system used on boats.

FUTURE APPLICATIONS OF SOLAR ELECTRICITY

The examples of practical applications of solar electricity given in this chapter are not complete. We have merely tried to give some idea of the many ways in which solar electricity is being used today.

The future applications of solar electric generators is almost unlimited. With each new technological advance the cost of solar cells drops making other applications economically feasible. In addition to the normal advance in the state-of-the-art, a great deal of effort is being expended toward completely automated production of solar cells that will further reduce the cost.

As we stated earlier, the goal of the U. S. Department of Energy is to reduce the cost of solar cells to less than $0.50 by 1986. Even if this goal isn't completely realized, the cost will continue to decline. This will make solar electricity the most economical approach to electric power in an ever increasing number of applications.

Fig. 6-17. Solar powered attic fan.

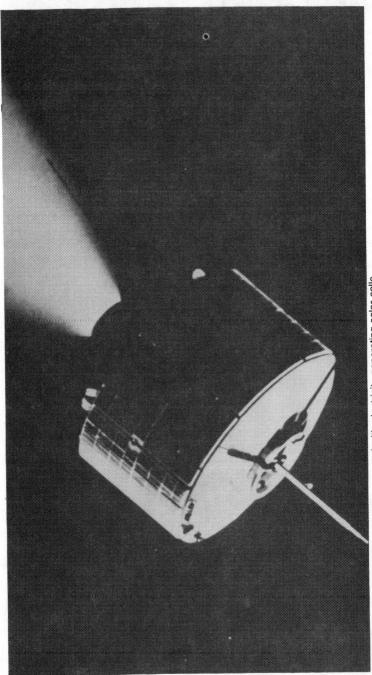

Fig. 6-18. A man-made space satellite covered with electricity-generating solar cells.

TWO-WAY COMMUNICATIONS

The use of solar cells to generate electricity at remote mountaintop locations to power two-way communications gear is expanding. Figure 6-19 shows a power-generating system in use atop a New Hampshire mountain, providing electricity from the sun for a two-way radio repeater.

On top of Mount Cardigan, west of Newfound Lake, a solar electric generator is being used to power a General Electric UHF solid-state radio repeater for the Department of Public Works and Highways of the State of New Hampshire.

The repeater gives line-of-sight transmission and reception for radio contact with units in the New Hampshire valleys. Radio signals from low-powered mobile and portable radios arrive at the repeater where they are boosted and rebroadcast over a wide area to other mobile, portable and base stations.

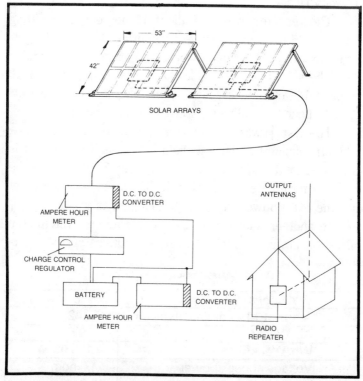

Fig. 6-19. Repeater pictorial block diagram.

Patrol vehicles use the repeater to communicate with six base stations and 50 other vehicles. The repeater is at an altitude of 3200 feet, atop the mountain, which lets it "see" down into many valleys.

The repeater uses no commercial electrical mains and provides radio cover of 800 square miles.

The power-from-the-sun generating system consists of three parts: a solar panel array, a Solarex Powermizer charge-control circuit, and storage batteries. The solar panels receive sunlight and generate electricity. That electricity is stored in the batteries which are under control of the Powermizer.

When the repeater is not actually retransmitting a signal, but is listening, it draws 300 mA of current from the batteries. When it is transmitting, it uses 24 amps of current. It runs on 12 volts DC.

The repeater averages about 15 percent transmitting time and 85 percent listening time during the normal eight-hour work day, five days a week. It operates year round.

The solar panels provide 200 watts peak power or 13 amps at 14 volts DC in full sunlight. The batteries are 1008 ampere-hour lead-calcium photovoltaic storage units.

The system was designed for heaviest use in winter time when the amount of daily sunlight is lowest. Extra power is produced in the summer so the Forest Service uses the repeater for fire detection, also. If the Forest Service did not use the extra power, it would be dumped as waste.

In another sun-powered communications system, the SNOTEL environmental data communications system oper-

Table 6-1. Characteristics: 435HP Unipanel.

Watts (peak)[1]	6
Amps at 14V nominal (typical)	.42
Ah/Week[2]	9.5
Wh/Week[2]	135.0
Voltage at peak power[3]	14.0 nom.
V_{OC}[3]	14.0 nom.

ated by Western Union for the Soil Conservation Service, U.S. Department of Agriculture, is on the air in Idaho.

SNOTEL includes 500 remote monitoring stations, each powered by a battery pack charged by solar panels. The stations are in remote locations where sunpower frees them of dependence on power lines. One site is near Moore's Creek Summit in Idaho, powered by a solar panel atop a tower.

SNOTEL relies on reflection of radio waves from electrons in meteor trails to transmit data from remote stations to base stations separated by as much as 1200 miles. Since meteor trails exist only from a few thousandths of one second to a few seconds of time, a "burst" type of transmission is used.

Each base station surveys the atmosphere for meteor trails capable of reflecting signals to remote sites. When a suitable path is found, the base sends a signal to one or more remotes which then are triggered to transmit back to the base whatever data the remote site has collected and stored. Each remote unit has 16 sensors of environmental data such as

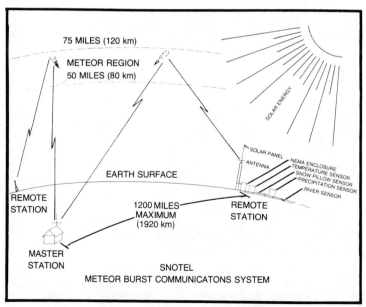

Fig. 6-20. SNOTEL meteor burst communications system.

snow depth, temperature, stream flow, and wind velocity and direction. Figure 6-20 shows the SNOTEL meteor burst communications system.

Typical remote sites use 20 mA of current at 12 volts DC on average. The solar panels and storage batteries are beefed up to provide enough backup for extended periods of inclement weather. Table 6-1 shows electrical characteristics of the Solarex Corp. solar panel used.

The 435HP Solarex Unipanel is made up of 36 silicon solar half-cells in series mounted on a fiberglass board in a special stabilized silicone rubber which provides a transparent, weather-resistant, and corrosion-free encapsulation. The cells feature a tantalum oxide anti-reflective coating and redundant collector pattern for high conversion efficiency, and corrosion-resistant electrodes. The shunt regulator is heat-sinked to avoid overheating, and each panel is carefully quality-controlled and tested in house. The 435HP measures 10-3/8″ × 12½″ × ¾″ and weighs 4½ pounds. Other tested applications of the 435 HP include remote telemetry units for oils and gas pipelines, supervisory control and alarm systems, navigation aids, and hydrological and meteorological data collection stations.

Chapter 7
Solar Electric Projects

The silicon solar cell is one of the most fascinating devices available to the educator or experimenter. All of the components required for solar electric experiments are readily available at low cost. The circuits involved are much simpler than most other electronic circuits. All solar cell experiments are educational and many result in useful devices.

The number of things that can be powered by solar electricity is limited almost entirely by the imagination and ingenuity of the experimenter. Many new electronic components operate on very small amounts of power and can be powered with only a few solar cells at most. In this chapter we will describe a few projects that can be performed with a minimum number of components. Projects range from demonstrations of solar energy to providing operating power for small devices.

DEMONSTRATION PROJECTS

Projects that demonstrate solar electricity are fascinating and are of value either educationally, esthetically, or merely as conversation pieces. Used in schools, they will stimulate students interest in science and increase their awareness of the importance of new energy sources.

An excellent device for demonstrating solar electricity is the solar fan cube shown in Fig. 7-1. This 3¾-inch cube is constructed of transparent plastic and contains a solar cell connected to a motor driven fan. Whenever the solar cell is illuminated, the fan will rotate, providing an excellent demonstration of the energy available in sunlight.

The motor and connecting wires are available in different colors to enhance the appearance of the device.

An even more attractive demonstration unit is the sun flower shown in Fig. 7-2. This device consists of a solar cell connected to an electric motor to which are attached the petals of a "flower." When sunlight hits the solar cell the motor will turn on its axis thus rotating the petals of the flower.

COMPONENTS

All of the components needed for solar cell experiments are readily available. Many popular sizes of solar cells, including Microgenerators, are now available in blister packs. An example of such a pack is shown in Fig. 7-3. The experimenter can select the size of cell best suited to the projects he has planned.

A word of caution is in order regarding the handling of solar cells. Although solar panels that are packaged for application are rugged and almost indestructible, individual solar cells are fragile devices. The average solar cell is very thin and can easily be broken with rough handling. The surface can become dirty thus reducing its energy conversion efficiency. As a general rule the solar cell should be handled just like any other delicate device such as a lens. Care should be taken not to subject it to shock or stress until it is mounted on some firm support. Once the solar cell is mounted on a solid surface, it will withstand ordinary handling with no adverse effects.

The individual solar cell usually doesn't have any leads attached to it, and the experimenter will have to solder leads. The positive lead is soldered to the back of the cell and the negative lead is soldered to the metallized surface on the front. Excess heat can very easily damage the cell. Only small wire,

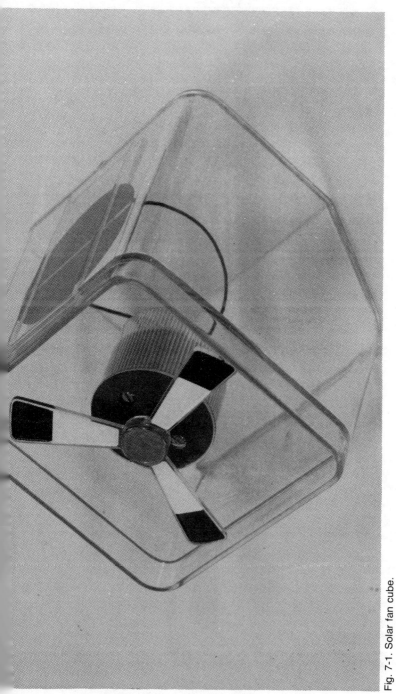

Fig. 7-1. Solar fan cube.

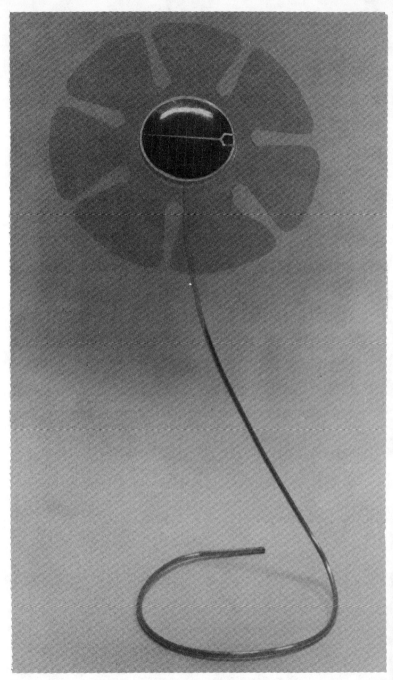

Fig. 7-2. "Sun Flower" demonstrates solar electricity.

AWG #26 or smaller, should be used. Use a very small soldering iron, and do not apply heat any longer than necessary to make a good connection.

As explained in earlier chapters of this book, the individual solar cells can be connected in series to get the desired voltage and in parallel to get the desired current. In most applications other than very small ones, cells will be connected in both series and parallel.

A very convenient source of components for many experiments and demonstrations is the Photovoltaic demonstration

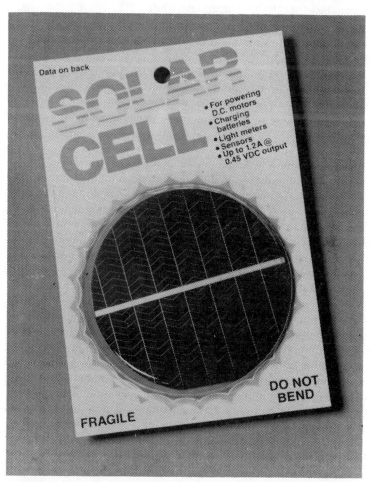

Fig. 7-3. Blister packed solar cell.

kit shown in Fig. 7-4. The kit contains nineteen different component parts that can be interconnected in many ways to provide hundreds of different circuits. The solar cells in the kit are packaged in modules so that they can be handled without the fear of damaging the cells. Electrical connections are already made to the cells so the problem of carefully soldering leads is avoided. This is particularly helpful in classroom situations where unprotected cells might be damaged by inexperienced students.

The kit contains a small motor and a buzzer that can be operated by the solar cells. An instruction book is provided that gives the characteristics of the components and circuits that can be built by the experimenter. The contents of the kit are as follows:

- 4 solar cell modules
- 1 electric motor module
- 1 buzzer module
- 1 rheostat module (10 to 100 ohms)
- 4 250-millimeter leads
- 500-millimeter leads
- 1-millimeter lead
- 1 ammeter module (0 to 1 ampere)
- 1 voltmeter module (0 to 5 volts)
- 1 rheostat module (0.5 to 5.0 ohms)

DIRECT SOLAR ELECTRIC POWER

There are many projects that the experimenter can build that will only operate when there is light available. In these experiments the solar cells are connected directly to the device to be operated without any storage batteries.

When using the solar cell without any storage battery, its characteristics as a source must be taken into consideration. The familiar sources of power for electronic circuits such as batteries and power supplies have a very low internal impedance. This is the same as saying that their terminal voltages tend to stay relatively constant while the current varies with the load. A solar cell is somewhat different. As we explained in

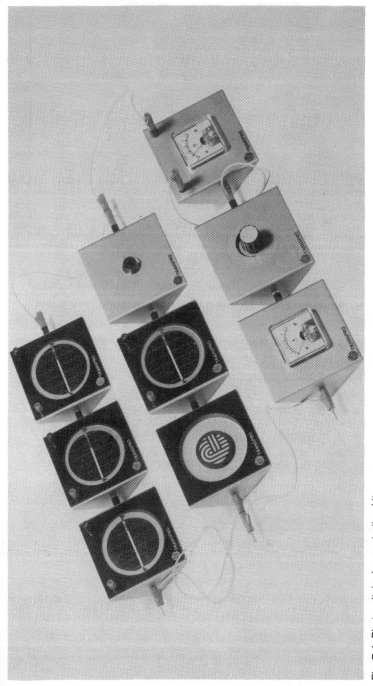

Fig. 7-4. Photovoltaic demonstration kit.

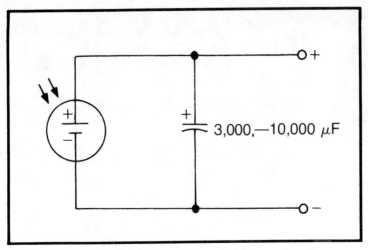

Fig. 7-5. Using a capacitor to regulate the voltage from a solar cell.

an earlier chapter, when the light is constant, the solar cell will look much more like a constant current source over a wide range of load variation.

Thus to apply a solar cell in powering something like a small transistor radio, we will need to do something to make it behave more like a constant voltage source. Fortunately, this is relatively easy. About all we have to do is to put a rather large capacitor across the output of the solar cell. The arrangement is shown in Fig. 7-5. The size of the capacitor is not critical, but in general, the larger the better. Because of the fact that solar cells have a low output voltage, the capacitors are relatively inexpensive.

Figure 7-6 shows the diagram of a solar electric generator that can be used to power a wide variety of electronic circuits including radio receivers, transmitters, and small instruments. The output voltage of the array is 5.85 volts and the available current depends on the size of the individual cells and the amount of light available. This particular voltage was chosen because most small transistor circuits will work well at 6 volts, but there is nothing magic about it. Any series combination of cells can be used to provide the desired terminal voltage for any particular application.

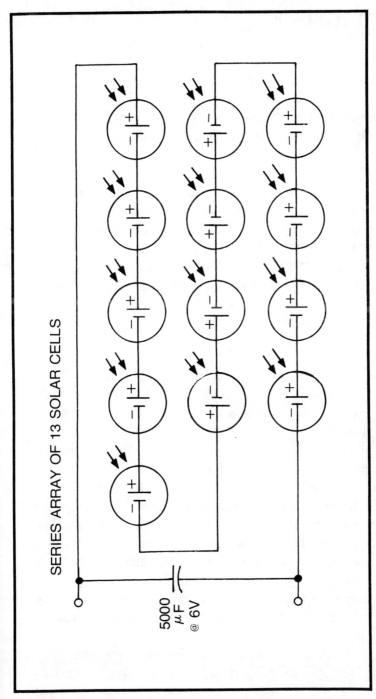

Fig. 7-6. Practical solar electric generator for small circuits.

BATTERY CHARGING

The big limitation of the direct solar electric generator described in the above paragraphs is that it will only furnish power when it is illuminated. Although this is acceptable for many applications, in the general case we want equipment to operate at any hour whether or not daylight is available. This is easily accomplished by including a storage battery of some type in the generator. The solar cell or cells can then be used to charge the battery when daylight is available.

Figure 7-7 shows a solar electric battery charger. The number of cells used in the solar array depends on the voltage and current required. The diode is included to assure that current will only flow in one direction and the battery will not discharge through the solar cells when there is no light. The resistor is provided to limit the charging current. In many small battery chargers, the resistor will not be required at all. Some sealed lead-acid batteries are designed to float across a charger at all times. In other cases, where not much current is drawn from the battery during the night, there may be a danger of the battery overcharging during the day.

The voltage output from the solar battery charger should be selected to provide enough charging current to recharge the battery during daylight hours. The techniques explained in Chapter 5 can be applied here. Usually a nominal voltage output of 14 volts works well for charging 12-volt batteries and 28 volts is satisfactory for charging 24-volt batteries.

When a battery charger is designed to recharge the batteries in an existing piece of equipment, the designer has little choice as to the voltage and current rating of the charger. These will be determined by the voltage and current ratings and the duty cycle of the device. However, if a piece of equipment is being designed specifically to be powered by solar electricity, the designer can do many things to simplify the power supply.

One thing that can be done when designing a device to operate from a solar electric supply is to design the circuit to minimize power consumption. If readouts are used, the liquid

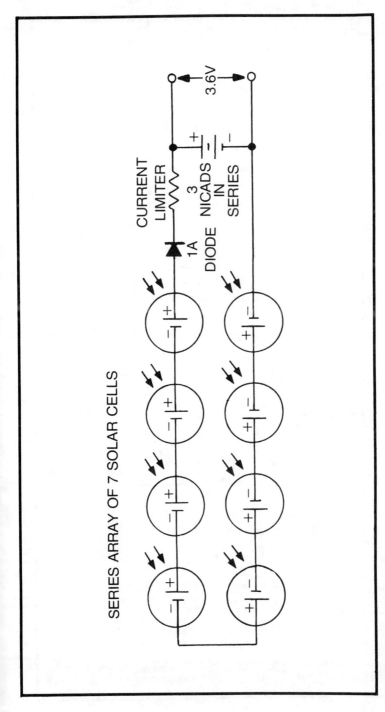

Fig. 7-7. Battery charger.

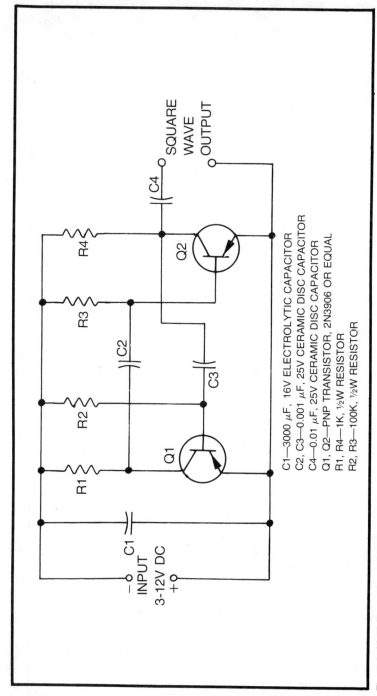

Fig. 7-8. Multivibrator inverts DC voltage from solar cell.

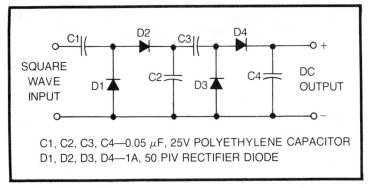

C1, C2, C3, C4—0.05 µF, 25V POLYETHYLENE CAPACITOR
D1, D2, D3, D4—1A, 50 PIV RECTIFIER DIODE

Fig. 7-9. Voltage multiplier.

crystal has a distinct advantage over light-emitting diodes as far as power consumption is concerned. Similarly, in digital circuits, CMOS technology consumes much less power than TTL chips. In analog circuits the "micropower" operational amplifiers have an advantage.

INCREASING VOLTAGE FROM SOLAR CELLS

In applying solar electricity to some small electronic projects, we run across the situation where two or three solar cells will provide enough power to operate the device in question, but will not furnish enough voltage. Rather than add more solar cells to the generator just to increase the voltage, it might be worthwhile to build a small inverter and voltage multiplier.

Figure 7-8 shows the circuit diagram of a small multivibrator that will operate on a very small supply voltage. This circuit uses germanium transistors because the base to emitter voltage across a germanium transistor is only about 0.2 volt, whereas the base to emitter voltage of a silicon transistor is about 0.7 volt.

The multivibrator is actually a square wave oscillator. It's output is a square wave alternating current waveform having a frequency of about 1000 hertz. Now that we have an alternating voltage, we can increase it by using a voltage multiplier.

Figure 7-9 shows the circuit of a voltage multiplier. This particular circuit will triple the voltage from the multivibrator,

Fig. 7-10. Electronics experimenters and hobbyists can fasten this breadboard-style solar-cell electricity generator to perfboards or other breadboard lashups.

but by adding sections we can get an increasing amount of voltage multiplication. Remember, that although the circuit will increase voltage, it cannot increase power. Therefore, every time that we increase the voltage, the circuit will furnish less current—P = IE.

The multivibrator shown in Fig. 7-8 will operate at a low voltage. It will always start at 3 volts and with a little careful trimming of the values of the various components, it can be made to operate at applied voltages as low as 1 volt.

Chapter 8
Accessories

Installation of a solar electric generator can be simplified considerably by the use of accessories that save the designer time and effort. These accessories include charge controllers, an ampere-hour meter, a tracking pedestal, and various items of mounting hardware.

CHARGE CONTROLLER

Whenever the ampere-hour output of a solar generator is more than about five percent of the ampere-hour rating of the storage batteries that are used with it, it is recommended that some form of charge control be employed. Otherwise, the batteries may be overcharged. At the best this may mean that the electrolyte will evaporate requiring refilling, and at worst it may damage the batteries.

The Solarex Powermizer® charge controller was developed specifically for use in solar electric systems (Fig. 8-1). It is fully compatible with all types of solar panels, as well as with various types of storage batteries including lead-acid, nickel-cadmium, and gelled electrolyte types. The charge controller has solid-state circuitry so it will tend to have a life as long as that of the solar cells.

Fig. 8-1. Powermizer charge controller. Developed specifically for solar electric generators, the Powermizer prevents overcharging of storage batteries.

Fig. 8-2. Ampere-hour meter and calibrated solar panel.

MEASUREMENT EQUIPMENT

Figure 8-2 shows an ampere-hour meter and a calibrated solar panel. The ampere-hour meter will indicate the actual ampere-hour output of the panel. This arrangement will give the best possible indication of the amount of solar energy available at any particular location. By leaving the calibrated solar panel in a given location for an extended period of time, we will get a factual measure of the amount of energy. The arrangement has the advantage that it requires no attention at all. Once installed, it will continuously monitor the ampere-hours produced by the calibrated panel.

TRACKERS

As we have mentioned earlier, the best orientation for a solar array in most parts of the United States is due South. Of course, this means that the array is not pointed directly at the sun all day long, but is pointed in the correct direction during the hours of maximum daylight. In instances where concentrator cells are used to get the maximum possible amount of

Fig. 8-3. Dual-axis tracker. This tracking device is weatherproof and mechanically simple, yet provides enough output and accurate two-axis tracking capability for solar-electric and thermal systems.

Fig. 8-4. A power module mounted on a tracker. The multicell sun sensor continually monitors the tracker's alignment with the sun. It uses solid-state circuitry to provide a voltage differential proportional to any tracking error.

energy from the sun in a limited space, it is helpful to keep the array of solar cells pointed directly at the sun at all times.

A tracking pedestal contains motors that will keep the array pointed at the sun at all times. A small sensor mounted on the array provides electrical signals that tell the control system which way to turn the array to get maximum sunlight.

Chapter 9
Non Power Photovoltaic Cell Applications

So far in this book we have looked at the silicon photovoltaic cell as a source of electrical power. Indeed, this is probably the most significant application of the photovoltaic effect. We must not forget, however, that the solar cell is a light sensitive device and therefore has many applications other than converting light into useful quantities of electrical energy. In many circuits, a solar cell can be used both to sense the level of light and to provide operating power for the circuit. In this chapter we will consider a few of the nonpower applications of solar cells.

LIGHT METER

Inasmuch as the solar cell responds to the light level, we can use it with a meter to measure or sense the level of light. In our study of the properties of the silicon solar cell we found that, except for very low light levels, the current output of the cell is very nearly proportional to the power density of the incident light. This means that to measure the output of the cell, we will do best if we measure the short-circuit current output of the cell. We can best do this with a microammeter or milliammeter. Fortunately, most multimeters have milliam-

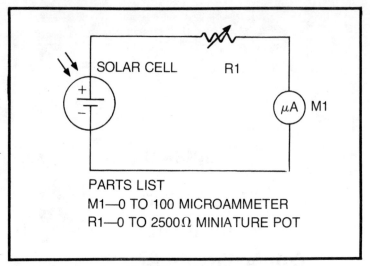

Fig. 9-1. Use of a solar cell in a light meter.

pere scales and many have microampere scales. We do not want to use a voltmeter, because with a constant load, the voltage output of a cell doesn't change very much with changing light levels.

Figure 9-1 shows the circuit of a simple light meter. Here a multimeter is used as an indicating instrument. The various current scales of the multimeter can be used to give different ranges of light level measurement.

Calibration of the light meter poses a few problems. Remember that although the resistance of an ammeter is very low, it is still not zero so we won't actually get the short-circuit current of the cell. The microampere ranges of some multimeters actually have a rather high resistance. The best way to calibrate a light meter is to use the services of a photometric laboratory. Some camera shops that repair exposure meters have equipment that can be used for calibration.

When nothing better is available, some rough calibration can be made by placing the solar cell right beside a photographic exposure meter and subjecting the pair to varying light levels. The light meter can then be at least roughly calibrated according to the indication of the exposure meter. Lacking any means at all of calibrating the light meter, you can always use it to give a relative indication of the available light.

A light meter can be used to demonstrate many interesting properties of light. For example, the meter can be placed in front of a light source and various materials inserted in the path of the light. It is interesting that objects that appear to be transparent, such as pieces of glass or plastic, will actually reduce the amount of light reaching the meter. Blowing smoke between the light source and the light meter will show how photovoltaic cells can be used in a smoke detector.

LIGHT OPERATED RELAYS

Another application of the solar cell is to operate a relay. If a sensitive relay is available, it may be operated directly by the output of the solar cell without any other power supply. A typical circuit is shown in Fig. 9-2. Most relays that are readily available will not operate on the output of a single solar cell. The reason is the resistance of the coil of the relay is usually too high. The current from the solar cell is great enough to operate the relay, but the voltage isn't high enough. There are

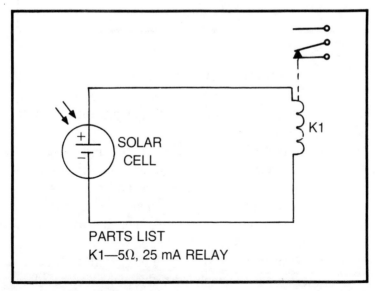

Fig. 9-2. Light sensitive relay.

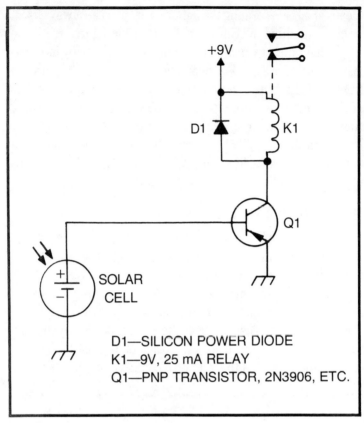

Fig. 9-3. Light operated relay with amplification.

two approaches to avoiding the cost of a low resistance relay. One is to simply use a few more solar cells in series to increase the voltage output. Another is for the experimenter to make his own relay.

Building a relay isn't difficult because of the availability of magnetic reed relays. These relays are available with coil forms so that the experimenter can wind his own coils. Furthermore a small permanent magnet may be placed near the relay to provide almost, but not quite, enough magnetic force to operate the relay. All the coil of the relay will then have to do is to provide enough additional force to operate the relay.

If an extra power supply isn't objectional, a larger less sensitive relay may be used with a transistor amplifier. Figure

Fig. 9-4. Examples of various solar cells manufactured by Solarex Corp. for use on Earth.

9-3 shows a typical circuit. Note that if a germanium transistor is used, the output of a single solar cell will operate the circuit. This is because the base-emitter voltage of the germanium transistor is much less than that of a silicon transistor.

The applications of a light operated relay are many. Note that we have used a single-pole, double-throw relay in the circuit of Fig. 9-3. Thus we can either turn something on or turn it off when the light level increases. A typical application is a circuit that turns off outside lights during daylight hours. Another application might be where a flashlight is used as a light source to control something or other.

Glossary of Solar Electric Terms

absorption. When light strikes a surface, some of it penetrates into the material and doesn't come out. This is called *absorption*.

angstrom. The angstrom is a measure of the wavelength of light. One angstrom = 10 nanometers = 10^{-10} meter.

AM-air mass or atmospheric mass. This is a measure of the absorption and scattering of light by the earth's atmosphere. AM 0 indicates that there is no atmosphere in the path of the light. This is the condition out of space where the power density of light is about 136 mW/cm^2. AM 1 is the condition at the surface of the earth where the light passes through one atmosphere. The power density is about 100 mW/cm^2.

ampere-hour meter. An instrument that monitors current with time. The indication is the product of current in amperes and time in hours.

blocking diode. A semiconductor connected in series with a solar cell or cells and a storage battery to keep the battery from discharging through the cell when there is no output, or low output, from the solar cell.

concentration ratio. The amount that light is magnified by a focusing system. For example, if a lens or reflector system increases the power sensitivity of sunlight from the normal 100 mW/cm^2 to 300 mW/cm^2—a magnification of three times—the concentration ratio is 1 to 3.

concentrator cell. A solar cell designed at light power densities much greater than the normal power density of sunlight at the surface of the earth. Concentrator cells can be used with focusing arrangements that increase the power density of sunlight hundreds of times.

efficiency of a solar cell. The ratio of the electrical power output of a solar cell to the solar power that it intercepts. For example, a solar cell 3 inches in diameter intercepts about 2.39 watts of solar power under full sun conditions. If the electrical output of this cell is 0.34 watt, the efficiency will be

$$\frac{0.34}{2.39} = 0.14, \text{ or } 14\%$$

full sun. The full sun condition is the amount of power density received at the surface of the earth at noon on a clear day—about 100 mW/cm^2. Lower levels of sunlight are often expressed as 0.5 sun or 0.1 sun. A figure of 0.5 sun means that the power density of the sunlight is one-half of that of a full sun.

insolation. The amount of sunlight reaching an area. Usually expressed in milliwatts per square centimeter.

incident light. The incident light is the amount of light reaching an object.

nanometer. A unit of measurement of the wavelength of light. One nanometer = 10^{-9} meter.

photon. In the quantum theory of light, light consists of tiny bundles of energy. One of these bundles is called a *photon*.

photovoltaic cell. A photovoltaic cell is one that generates electrical energy when light falls on it. This term distinguishes it from photoconductive cell (photoresistor), which changes its electrical resistance when light falls on it.

powermizer. A trade name of the Solarex Corporation for a battery charge controller that can be used with solar cells to prevent overcharging of a battery.

quantum yield. A technical parameter from quantum physics. For practical purposes the quantum yield of a solar cell is the spectral response of the cell, that is, the relative output at different wavelengths of light.

reflected light. When light reaches an object some of it may be absorbed, and some may be bounced off the surface. The light which is bounced off the surface is called *reflected light*.

solar cell. A photovoltaic cell that is designed specifically to produce electrical energy from sunlight.

solar constant. In physics the solar constant is the rate at which energy is received from the sun at just outside the earth's atmosphere at an average distance of the earth from the sun. This is about 136 mW/cm^2.

solar energizer. A trade name of the Solarex Corporation for a solar panel.

solar panel. A collection of solar cells connected in series, in parallel, or in series-parallel combination to provide greater voltage, current, or power than can be furnished by a single solar cell. Solar panels can be provided to furnish any desired voltage, current, or power. They are made up as a complete assembly. Larger collections of solar panels are usually called solar arrays.

temperature coefficient of a solar cell. The amount by which the voltage, current, or power from a solar cell will change with changes in the temperature of the cell.

Unipanel. A trade name of the Solarex Corporation for a type of solar panel.

voltage-current characteristic. A plot of the current output of a solar cell versus its terminal voltage. Usually several curves are given for different amounts of sunlight reaching the cell.

Index

A

Agricultural applications of solar electric power	79
Ampere-hours	62
Antireflection coating	38
Applications of concentrator cells	94
Applications of solar cells	75
Array of solar panels	33, 73
Arrays	35

B

Batteries	66
ampere-hour capacity	70
specifications	69
current requirement	69
operating temperature	70
size and weight	70
voltage requirement	69
Battery charger	119
Battery charging	118
Battery field	71
Blister packed solar cell	113
Boron doping	37

C

Capacitor	116
Charge controller	125
Color of light	11
Complete solar arrary	86, 87
Complete solar electric generator	71, 72
Concentrator cells	50, 51
Conversion factors for energy	12
Conversion factors for power	13
Conversion factors for power density	13
Conversion factors for wavelengths	15
Current curves	25

D

Demonstration projects	109
Direct solar electric power	114
Doping	20

E

Efficiency percentage	37
Electrical specifications	41
Electric connection pattern	51
Electrolytic corrosion	88, 90, 92
Energy	11
Energy conversion efficiency	32
Energy source	30
Equations of solar cell voltage	41

F

Fiber optics	78
Fiber optics repeater	81

FM translator 76
Future applications of solar electricity 102

G

Germanium 19

H

High density solar panels 46
Hole-electron pair 21

I

Increasing voltage from solar cells 121
Intermittent loads 62

J

Joules 12

K

Kilogram-calories 12
Kilowatt-hours 12

L

Langley 12
Light meter 131
Light operated relays 133
Light sensitive relay 133
Load requirements 62
Low concentrator array 54

M

Maximum power 27
Measurement equipment 128
Microgenerators 55, 58
Microgenerator applications 97
Microgenerator array 56
Microgenerator battery charging circuit 97
Military applications 94
Multivibrator 120

N

Nanometers 14
Non power photovoltaic cell applications 131
N-type materiald 20

O

Open-circuit voltage 26
Output power 27

P

Parallel connections 42
Peak sun hours over four week period 60, 63
Peak sun hours per day 60
Photons 17
Photosensitive surface 37
Phosphorus 37
Phosphorus doping 37
Photovoltaic band 33
Photovoltaic cell 21, 37
Photovoltaic demonstration kit 115
Photovoltaic effect 19
PN junction 22
Power density 11
Power input 37
Power output 37
Principal of focusing 51
P-type material 20

R

Recombination 21
Regulated microgenerator 56

S

Semiconductor cells 47
Semiconductors 19
Series connections 42
Silicon 19
Silicon solar cell 25, 38
SNOTEL meteor burst communications system 107
Solar cell 9, 30, 37
Solar cell configurations 36
Solar cell efficiency 9, 37
Solar concentrators 95
Solar electric cathodic protection 88, 93
Solar electric installation 60
Solar electric power system 59
Solar electric projects 109
Solar cell experiment components 110
Solar cell specifications 40, 42
Solar cell output 31
Solar electric generators 35, 117
Solar electric sunflower 112
Solar energy 10
Solar fan cube 111
Solarex Corporation 10
Solarex powermizer 125, 126
Solar microgenerators 52
Solar panels 44
Solar power for recreation 100
Solar powered attic fan 99, 103

Solar powered calculator	99
Solar powered flashlight	99
Solar power in remote areas	84
Solar power in the construction industry	83
Solar powered irrigation system	82
Solar powered radio repeater	77
Solar powered road construction flasher	85
Solarvoltaic panel	48
Spectral response	30, 31
Standard microgenerator	56
Storage battery characteristics	72
Sunlight	11
Sunlight available	59
System calculations	64

T

Temperature characteristics of solar cell	39
Temperature influence	38
Trackers	128
Transmission signals of radio and television	76
TV translator	76
Two-way communications	105
Types of batteries	67

U

Unipanel	46
Unipanel characteristics	106
Units of wavelength	14

V

Voltage-current curve	27
Voltage curves	25
Voltage multiplier	121

W

Wavelength	11
Watt-hours	12
Wavelength of light	11

Y

Yearly average peak sun hours	61

DISTRIBUTORS—NATIONAL DISTRIBUTION

EDMUND SCIENTIFIC COMPANY
Dept. SGSE
101 E. Gloucester Pike
Barrington, New Jersey 08007
Tel. 609-547-3488

ALLIED ELECTRONICS
Dept. SGSE
401 E. 8th Street
Fort Worth, Texas 76102
Tel. 817-336-5407

NEWARK ELECTRONICS
Dept. SGSE
500 N. Pulaski Road
Chicago, Illinois 60624
Tel. 312-638-4411

SOLAR INTERNATIONAL
Dept. SGSE
124 Rt. #3 North
P.O. Box 274
Millersville, Maryland 21108
Tel. 301-987-9666